Beginner's Guide to Electronics

Newnes books of allied interest

A BEGINNER'S GUIDE TO RADIO

A BEGINNER'S GUIDE TO TELEVISION

ELECTRONICS POCKET BOOK
edited by J. P. Hawker and J. A. Reddihough

DICTIONARY OF ELECTRONICS
by Harley Carter, A.M.I.E.E.

RADIO SERVICING POCKET BOOK
edited by J. P. Hawker

TELEVISION ENGINEERS' POCKET BOOK
edited by J. P. Hawker

HIGH FIDELITY POCKET BOOK
by W. E. Pannett, A.M.I.E.E.

INTRODUCTION TO HI-FI
by Clement Brown

'PRACTICAL WIRELESS'
RADIO AND TELEVISION REFERENCE DATA
compiled by J. P. Hawker

RADIO ENGINEERING FORMULAE AND CALCULATIONS
by W. E. Pannett, A.M.I.E.E.

By the same Author

INTRODUCTION TO MICROWAVE SPECTROSCOPY

Beginner's Guide
to
Electronics

by

TERENCE L. SQUIRES

A.M.Brit.I.R.E.

GEORGE NEWNES LIMITED

SOUTHAMPTON STREET

LONDON, W.C.2

First published 1964

Printed in Great Britain by Butler & Tanner Ltd, Frome and London

CONTENTS

PREFACE

Electronics now plays an increasingly important part in all walks of life. Electronic techniques are the basis of most modern scientific research, they make possible much of our everyday entertainment, enable industrial processes from the simplest to the most elaborate to be automatically controlled, play a vital part in air and sea navigation, and are essential to modern defence and warfare. Such a subject is bound to excite the interest of intelligent young people of a practical or scientific turn of mind.

This book has been written to help those who are thinking of starting a career in electronics. It assumes no prior technical knowledge on the part of the reader. The subject is dealt with without recourse to mathematics, emphasis being placed on illustrative diagrams so that an understanding of this complex subject can be rapidly gained.

In the first chapters the nature of electric currents, pulses and waveforms are described, and the basic concepts of resistance, capacitance, potential difference and so on explained. The basic components that make electronic engineering possible, such as the various types of valves, transistors and semiconductor devices, are then described and their operation explained. After this the purposes to which they are put, for example amplification and waveform generation, are described. The principles of electronic testing and measuring are then outlined. Later chapters survey the main branches of electronics—radio and television, medical electronics, radar, space research, industrial electronics and computers—with emphasis on the basic techniques used in each field. A chapter is devoted to a review of some of the subjects and developments that will play an important part in the future of electronics, and the final chapter describes the various ways of training to be an electronics engineer.

Terence L. Squires

I

ELECTRIC CURRENTS

The story of electronics began in 1883 when the American physicist and engineer Thomas A. Edison discovered that under certain conditions electricity will flow through a vacuum. He had been experimenting with a small metal plate placed inside an evacuated electric light bulb. When he made this plate electrically positive with respect to the filament of the bulb, he found that an electric current flowed through the vacuum between the filament and the plate, that is through the lamp. But before we consider the significance of this simple experiment, let us consider what an electron is and the laws it obeys.

When we talk about matter we usually mean nearly everything that we can detect by our senses. A simple definition of matter would be anything that occupies space and has weight. Matter can exist in any of three conditions: solid (or frozen), liquid or gaseous. Examples of these conditions (at normal temperature and pressure) are iron, water and air. At normal room temperature iron is frozen, that is solid, but if the temperature is raised to 1,000° C then it becomes molten or liquid. At room temperature water is molten or liquid whereas at 0° C it becomes frozen or solid. At nearly 200° C below zero air becomes liquid and at a few degrees lower still it becomes solid.

Generally speaking gases can be made into liquids or solids by changing their temperature and/or pressure, and liquids can be made gaseous or solid. In the case of solid matter, however, there are many complex types which can only exist as solids.

Molecules

If matter is broken down by chemical techniques it will be found to consist of molecules. For example, wood consists of molecules of resin and cellulose. Air consists of molecules of oxygen, nitrogen, carbon, hydrogen and a few rarer gases.

Molecules are in turn composed of one or more of the atoms of the hundred or so elements known to science. Most of these elements may be found in any of the three states previously mentioned. The element oxygen's atoms form up in twos to produce a simple oxygen molecule which is a gas. Iron and copper molecules are simply atoms of these elements held in a crystal structure forming a solid. Polytetrafluorethylene, on the other hand, a solid much used as an insulator in electronics, is a complicated molecule made up of atoms of the elements carbon, hydrogen and fluorine.

Simple molecules are common salt, which consists of only two atoms, one each of sodium and chlorine, and copper in which the molecule is just one atom of copper. In the case of copper the molecule and the atom are the same unit. A molecule of water (chemical abbreviation H_2O) consists of two atoms of hydrogen (H_2) and an atom of oxygen (O).

Electrons

An atom is the smallest unit which still retains the identity of an element. If the atom is broken down still further one obtains a collection of electrons, protons and other particles, many of which have only recently been discovered by scientists.

From the electronics point of view only two particles need concern the reader: electrons, and the remainder of the atom which is called the nucleus.

Of the hundred or so known elements, the hydrogen atom has the simplest structure, consisting of a single electron moving in orbit around a nucleus which is a single proton. This is illustrated in Fig. 1.

Engineers regard a 'thou', that is a thousandth of an inch, as being a very small unit of measurement and precision work is generally measured in 'thous' or, as they are sometimes called, 'mils'. The hydrogen atom is about one millionth of four 'thou' in diameter and the electron is about one ten-thousandth of the size of the whole atom. If the atom was scaled up so that the nucleus was equal to ten 'thou' in diameter (or 0·01 in.), then the electron would be in an orbit about 100 ft away from the nucleus. Thus one can see that the atom is mainly composed of space.

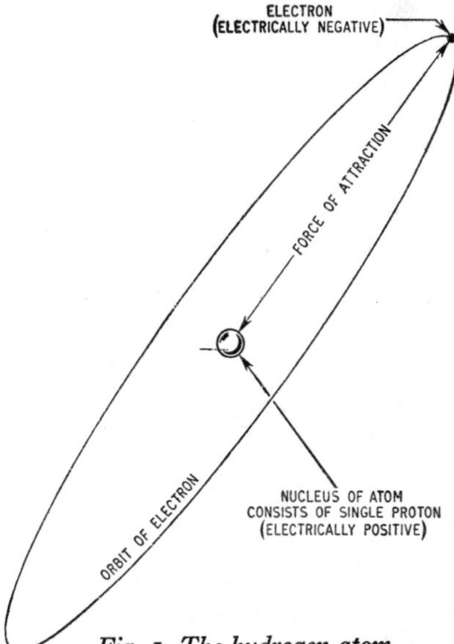

Fig. 1. The hydrogen atom.

Electricity

The electron has a negative charge: it is always attracted to any body which has a positive charge. In the nucleus of an atom are particles called protons which have a positive charge, and in normal circumstances this positive charge is equal to the negative charge of the atom's electrons. If an atom has fifty (negative) electrons then there will be fifty (positive) protons in the nucleus so that the electrical charges balance, for in its normal state the atom is always electrically neutral. If an atom loses an electron it has an excess of positive electricity and is called a *positive ion.* If it gains an electron it becomes a *negative ion.* The movement of electrons and ions under various influences forms the basis of all work in electronics.

Current

An electric current is the organized movement of electrons in a material, the number of electrons flowing past a certain point in a given time being the rate of current flow. Electric current (symbol *I*) is measured in amperes, milliamperes (1/1000 of an ampere) or microamperes (one millionth of an ampere), the last two units being the ones most commonly used in electronics. A microampere is equal to about six billion electrons flowing past a point in a second.

The actual number of electrons that flow—irrespective of time—is measured in coulombs. A 'box' containing a coulomb of electricity would store over six times a million, million, million electrons.

Conventional current

Like many other sciences, the study of electricity began to yield practical results before a basic theory to explain electrical phenomena could be formulated. Early investigators thought that electric currents flowed from the positive terminal of a source of electrical energy, for example a battery, to the negative terminal. This has come to be called the direction of *conventional current flow*. Today we know that the movement of the electrons, i.e. the current flow, is from the negative to the positive terminal.

The electronic engineer has always to remember this and when talking about the direction of the current he might be asked to specify whether he is talking about *electron current* or *conventional current*.

Conductors and insulators

Electric currents will pass more readily through some materials than others. Materials which readily pass electric currents are termed conductors while those that resist the passage of an electric current are known as insulators.

The atoms of some elements possess electrons which rotate in orbits forming an outer shell around the nucleus. Copper is an example, and is a commonly used conductor. Its real significance as a conductor, however, is its behaviour when a large number of atoms of it are grouped together to form a small block of the material. The atoms form a matrix or regular structure and remain linked in this fashion until the material is destroyed by physical or chemical means. Whilst the atoms remain fixed in a matrix, as shown in Fig. 2, they become ionized; that is, some of their outer

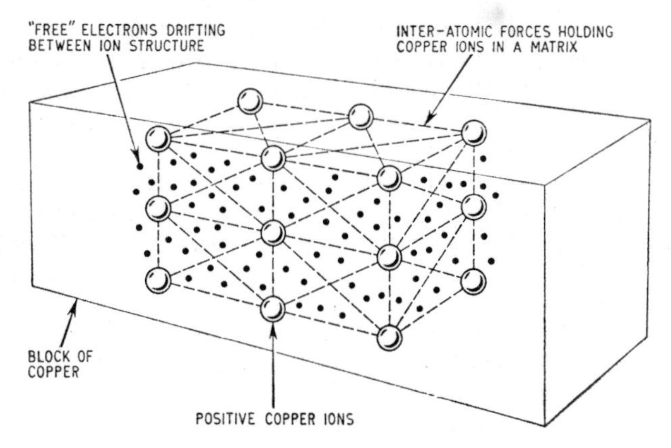

Fig. 2. Arrangement of atoms in a copper conductor.

Table 1. Conductivity of Various Materials

Materials that conduct electricity	*Materials that resist the flow of current but still allow some to occur*	*Materials that are good insulators and will not allow any current to flow through them*
Silver	Iron	Rubber
Gold	Oxides of metals	Bakelite
Copper	Carbon	Polytetrafluorethylene
Brass	Nickel-chrome	Polystyrene
Platinum	Germanium*	Ebony
Tin	Silicon*	Wood
Lead	Lead sulphide	Dry paper
Niobium		Ceramic
Palladium		Pure water
Molybdenum		Mica
Saline solution (salt water)		Asbestos
		Air

* These elements occupy a special category known as semiconductors.

Note that many conductors occur naturally as elements whereas many insulators are complex man-made structures.

shell electrons detach themselves from the parent atom and drift about the lattice rather like pigeons amongst the ornamentation on city buildings. Under the right influence, these free electrons will form an electric current. Most common metals act as conductors.

There are, on the other hand, elements which have atoms whose electrons orbiting on inner shells are more tightly bound to the nucleus and so are less ready to move about in the molecular structure. These materials are insulators, examples being pure water, dry paper, rubber, etc.

An electric current flows easily in a conductor, it flows with some difficulty in several 'in-between' materials, and negligibly or not at all in an insulator.

Conductors vary to some extent in their 'willingness' to conduct, for example, silver allows a greater current to flow under the same conditions than copper or iron.

Why an electric current flows in a conductor—an e.m.f.

An electron current is formed by the migration of electrons under some external influence such as a battery as shown in Fig. 3.

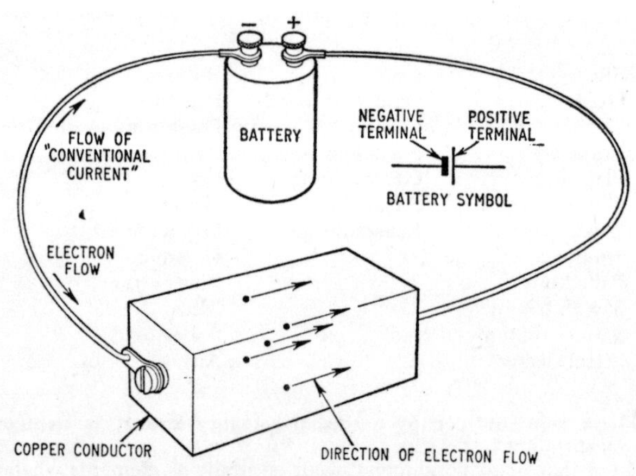

Fig. 3. Current flow in a conductor.

But what happens inside the battery? Inside the battery the circuit is completed by a paste or liquid electrolyte and two probes as shown in Fig. 4. When, and not until, the terminals outside the battery are shorted by an external conducting material, a chemical reaction occurs within the battery which produces ions, both positive and negative. (An ion is an atom or group of atoms having an excess or deficiency of electrons.) The negative ions, called anions, drift to the terminal of the battery which is marked 'nega-

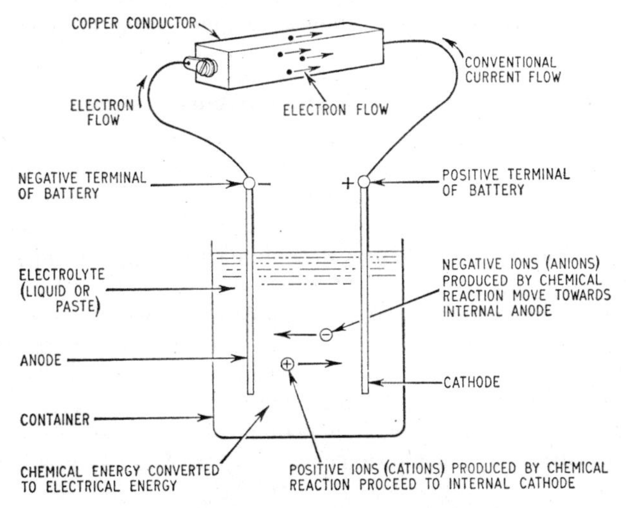

Fig. 4. Producing an e.m.f. from a battery.

tive' externally (although inside the cell it is behaving as a positive anode to attract the negative ions), whilst the positive ions—cations—are attracted to the terminal marked 'positive'. At the anode probe the anions give up their 'extra' electron, becoming neutral again. From the negative terminal the electrons given up by the anions flow in the external, conducting circuit towards the positive battery terminal. They then pass down the cathode probe and into the electrolyte to neutralize the cations—positive ions—which have migrated there as a result of the chemical reaction. In a battery, unlike a block of conducting material in which only electrons move, the ions are free to move. Inside the battery chemical energy is being converted to electrical energy and this provides an electromotive force (abbreviated e.m.f.) which drives the current around the circuit.

This electromotive force is measured in a unit called a volt. In electronics the millivolt (one thousandth of a volt, abbreviation mV) and the microvolt (one millionth, abbreviation μV) are the most common units to be employed. The single-cell battery just described is capable of providing an output voltage of about 1·5 volts.

Relationship between e.m.f. and current—resistance

It is e.m.f. that makes the electrons flow in a conductor. How many electrons flow in a particular conductor, that is, how large the current is, depends first upon how great is the e.m.f. and secondly on the resistance to current flow offered by the conductor. For a conductor of a given resistance, doubling the e.m.f. doubles the current flowing.

If we make a conductor of, for example, a mixture of copper dust and graphite powder the current that will flow will be reduced below the amount that would flow when pure copper is used even though the e.m.f. is kept at the same value. This is because the second material, graphite, is a poorer conductor—it resists the flow of current by not allowing enough electrons to take part in the action.

In practice all conductors have some resistance. To have zero resistance a conductor would have to be reduced in temperature to absolute zero (273° C below freezing point of water). This has nearly been achieved by using materials like lead and niobium at temperatures well below minus 100° C and minus 200° C respectively.

The symbol for resistance is shown in Fig. 5. A good conductor can be regarded as a theoretically perfectly-conducting wire with a hypothetical resistance in series, although this resistance will have a very small value.

Resistance (R) is measured in ohms (symbol Ω), but in electronics

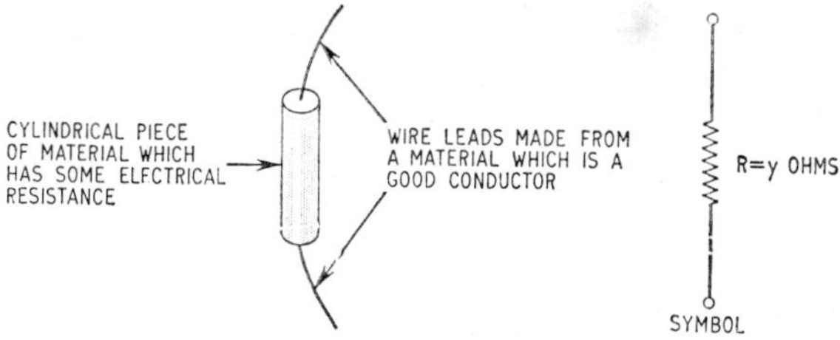

Fig. 5. Appearance of a typical small resistor and the symbol used in circuit diagrams to denote a resistor.

we may often be dealing with megohms (millions of ohms, MΩ) and kilo-ohms (thousands of ohms, kΩ).

To summarize, if we increase the e.m.f. we increase the current flow, if we increase the resistance we decrease the current flow.

Combinations of resistors

If resistors are connected together in parallel the effective resistance is reduced, while if they are connected together in series

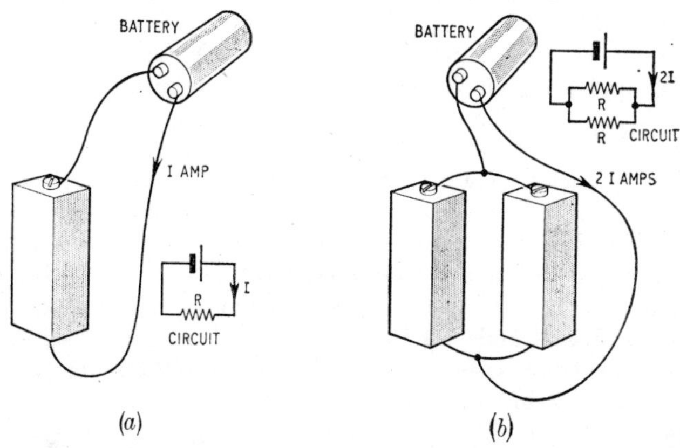

Fig. 6. Effect of parallel paths on current flow. (a) A single path through one conductor which has a small amount of resistance, R ohms. (b) The effect of two conducting paths on the current flowing. Both conductors have the same resistance value (R ohms) as the conductor in case (a), thereby allowing twice the amount of current to flow. (Two resistors in parallel means a smaller total resistance.)

the resistance is increased. This can be illustrated by an analogy. It is easier to empty a football stadium if you have many exits and it is easier for the e.m.f. to produce a larger current in a circuit if it has a number of parallel paths to drive the current through as seen in Fig. 6. Conversely, if the stadium exits were all placed in series it might take a whole weekend to empty it and, in the same way, if resistors are placed in series the total resistance presented to or 'seen' by the e.m.f. is greater and the resultant current flow is less. By the same reasoning if a conductor, say a wire x feet long, has a resistance of y ohms and its length is increased to $2x$ feet then the resistance would be increased to $2y$ ohms. If E volts were forcing a current of I amperes through the first wire, the same e.m.f. in the second case would produce only $I/2$ amperes.

Sources of e.m.f.

There are of course other sources of e.m.f. beside batteries. As everyone knows, coal is burnt to produce steam and mechanical power which will drive a large generator, the electrical output of which is widely distributed to various users. In places distant from such a mains supply a petrol-driven or wind-driven motor dynamo can be employed. (The Russians are using wind generators mounted on balloons six miles above the earth in what is known as the tropopause layer.)

Today, nuclear power is being used as a source of electrical energy. The coolant used to carry away exothermic heat from a nuclear reaction can be used instead of coal to produce steam.

Ten years ago batteries would have received scant attention in an electronics textbook since their use was limited to a few portable radio receivers. Today, however, with the immense growth of electronics and the wide use of transistor devices, the battery is once more becoming important as a source of e.m.f.

Batteries are of two kinds: primary cells and secondary cells. The former exhaust themselves and have to be thrown away after some time in use. Secondary cells can be recharged, after they have exhausted their chemical energy, by reversing the direction of the current through them as shown in Fig. 7.

Recharging could be done by means of another fully-charged

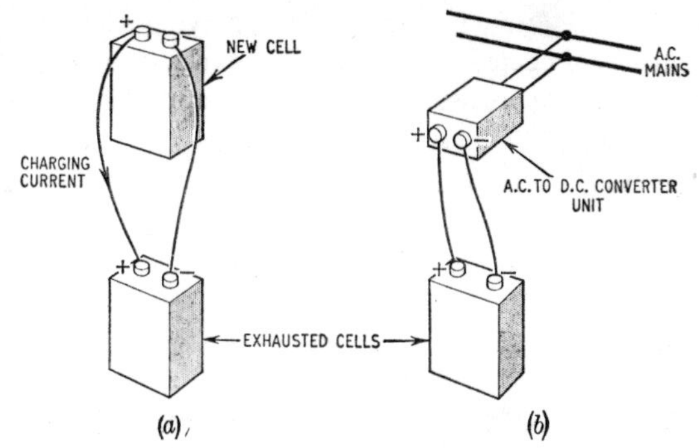

Fig. 7. Recharging an exhausted battery (secondary cell type). Method (a), using a new battery; method (b), normal method of recharging from the mains supply.

battery but in practice it is more likely to be done from a source derived from mechanical energy, such as the mains power supply.

Life of a battery

A battery supplies an e.m.f. at a value fixed by the chemical and physical characteristics of the cell. This e.m.f. will drive a current through an external circuit for a period of time which also is determined by these characteristics. The capacity of a battery is calculated in ampere-hours. For example, a 10 ampere-hour battery will supply, at a fixed e.m.f., a current of 10 amperes for 1 hour or 10 milli-amperes for about 40 days (1,000 hours).

A battery the size of an aspirin has been produced for use in spacecraft. It has an e.m.f. of 1·3V and a capacity of 36 milliampere-hours (1mA for 36 hours or 36mA for one hour).

Whatever the source of the e.m.f., it is apparent that in order to get electrical energy some other form of energy, mechanical or chemical, has to be drawn upon.

Potential difference

The amount of the e.m.f. present between the positive and negative terminals of a source of e.m.f. is termed the potential difference (p.d.). Like e.m.f., p.d. is expressed in volts. In Fig. 8

an e.m.f. derived from a battery is shown driving a current through two resistors connected in series. The e.m.f. can be said to be 'shared' by the two resistors, some of it 'appearing across' each

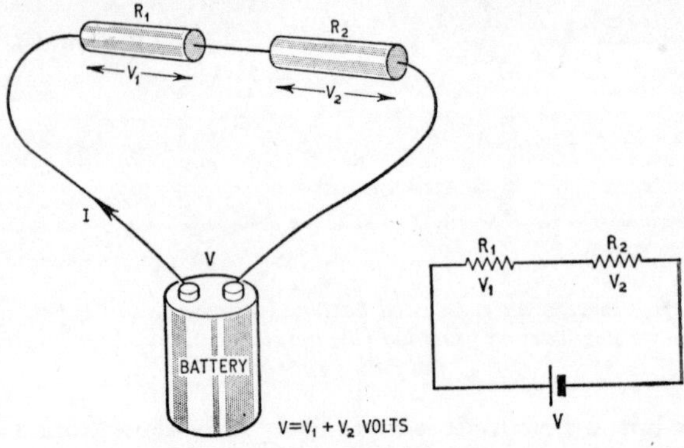

Fig. 8. Potential difference.

one. The amount that appears across each is known as the potential difference across that resistor.

If one of the resistors has a large resistance value and the other a small value most of the e.m.f. will appear across the larger value resistor, a proportionately smaller amount appearing across the smaller value resistor.

Ohm's Law

The mathematical relationship between potential difference, current flow and the resistance in an electric circuit was first investigated by the scientist George Ohm and is called Ohm's Law after him (as also is the unit of resistance, the ohm). This relationship is

$$R = \frac{V}{I}$$

where R is the resistance in ohms, V the potential difference (volts) and I the current (in amperes). The formula may of course also be written $I = V/R$, or $V = RI$. Thus if we know two of these quantities we can always calculate the third one.

Power

When a poor conductor or 'near-insulator', for example nickel-chrome wire, is placed across the terminals of a battery a certain small current will flow. This may be say one hundredth of the current that would have flowed if the conductor had been a very good one such as a block of copper. To produce the same current in both of these cases, a larger e.m.f. would be needed for the nickel-chrome wire. This could be obtained by using a number of batteries connected in series. To produce a high current flow through a greater resistance a high e.m.f. is required and this means that more batteries are needed and therefore more chemical energy is converted into electrical energy. The rate at which this energy is used up is called the *power consumption* of the resisting element and is expressed in watts (abbreviation W) or milliwatts (one thousandth of a watt, abbreviation mW).

Where does the energy disappear when a large current is forced through an electrical resistance? In poor conductors electron collisions occur in the molecular lattice structure and energy is expended in this process. If the nickel-chrome wire is felt when a current is passed through it, it will be noticed that the wire is warm. The electrical energy is being converted to heat-energy during the passage of current through the resisting material. This phenomenon is of course what happens in the ordinary electric light bulb and electric fire, where their resistance and the current forced through them make the filaments in both cases radiate light and heat energy.

In an electric fire thousands of watts are dissipated in this way. By way of comparison, the input circuit of a special measuring instrument used in electronics may dissipate only a few microwatts (millionths of a watt).

Output voltage and internal resistance

A very important idea which is constantly in the mind of the practising electronics engineer is that of internal resistance. In Fig. 4 a simple battery was shown forcing a current through a conducting block of copper. The current flow would be fairly high under these circumstances but, although the copper block has negligible resistance, the battery has an 'internal resistance' which limits the current flow. The output voltage provided by the battery is also affected by this internal resistance, i.e. some of the e.m.f. is lost as a voltage of potential difference across the battery's internal resistance.

All sources of e.m.f. have an internal resistance—sometimes called the source resistance. In the case of a battery it can be seen from Fig. 4 that the current flowing in the external circuit also flows inside the battery—in fact it is the same current. The probes and the electrolyte will have some resistance, even though it may be small—less than an ohm. Thus when the battery delivers a current, some of the e.m.f. is lost as potential difference across this internal resistance. The higher the current flow, the greater the e.m.f. lost and the smaller the voltage appearing across the terminals. This is demonstrated in Fig. 9.

Whatever the source of e.m.f.—motor generator, oscillator-

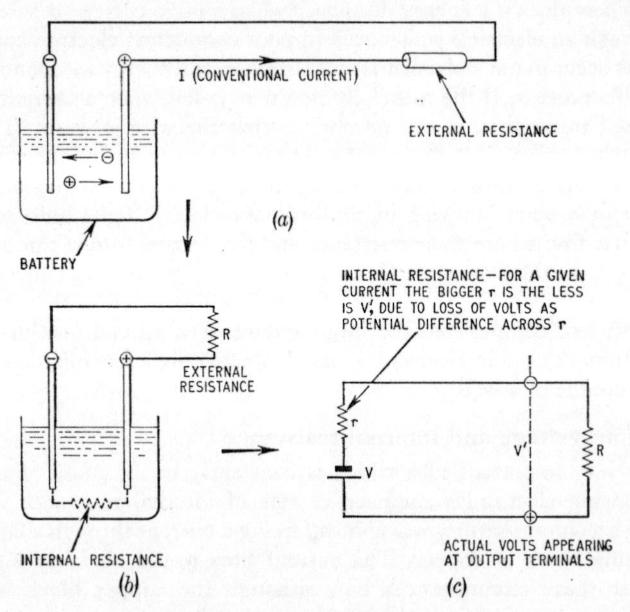

Fig. 9. The concept of internal resistance.

power source or rectifying power supply—each has this 'built-in' resistance. In some cases it is high: this can be a nuisance or, on the other hand, a protection. In other cases it is very low so that large currents can be drawn from the unit whilst the output voltage remains constant. (The higher the internal resistance the greater the voltage fluctuation with changes in the current flowing—see Ohm's Law, given earlier in this chapter.)

The generators that supply a town with electricity must have a very low internal resistance since they have to supply very heavy currents. Stabilized power supplies, now so commonly used in electronics, are also designed to have very low internal resistance because it is important to have a constant output voltage whatever value of current is drawn.

The concept of internal resistance—or as it is called in later chapters input or output impedance—is one of the most important in electronics. It is a factor which affects the design of many instruments and determines methods of measurement. For example, it is no use trying to measure accurately the output voltage of a circuit with a high internal resistance or impedance using a voltmeter which needs a considerable current to operate the movement. Were such a meter used, the output voltage of the circuit would fall as soon as the meter was placed across the points to be measured and thus a wrong reading would be obtained. In a case like this a special measuring instrument—a valve-voltmeter—can be used to obtain an accurate reading of the circuit output voltage.

Matching

Another important concept which is always occurring in electronics is that of correct matching (see Fig. 10). Consider the simple case of a generator supplying power to an electric light bulb. The generator and lamp are said to be *matched* when the optimum amount of power is being transferred from the generator to the lamp. This does not mean that all the power produced by the generator is delivered to the load (i.e. the bulb). Some, which is dissipated in the internal resistance of the generator, is wasted, only the remainder of the power being delivered to the load. It is, therefore, desirable to obtain conditions whereby the maximum amount of power is transferred to the load. It can be proved algebraically and readily shown with simple laboratory equipment that these conditions obtain when the resistance of the load is equal to the internal resistance of the generator. Under these conditions the generator is said to be matched to the load.

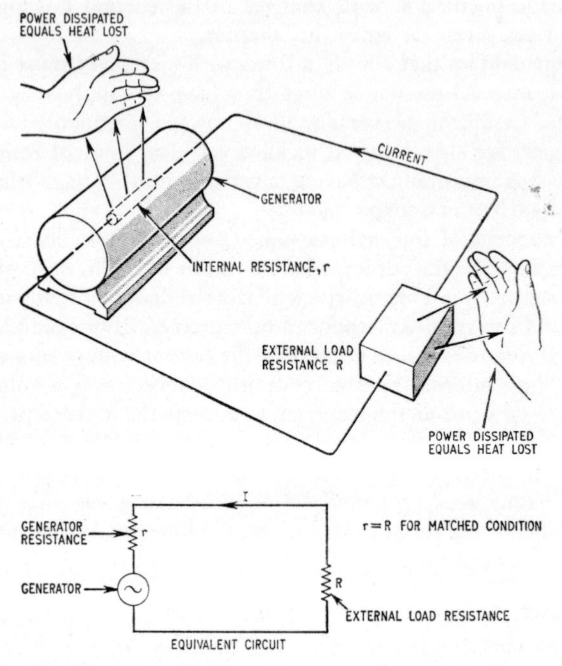

Fig. 10. The matched condition. When the generator and the external load are matched, hand detects thee sam heat from both.

The expression 'matched' occurs frequently in electronics. For example, an oscillator oscillating at say 10 million cycles per second (10 Mc/s) is really a generator and it gives its best performance if its output impedance (i.e. internal resistance) is matched to the impedance of the load. Again, an audio amplifier operating at say 1000 c/s will deliver maximum power to a loudspeaker if the loudspeaker's resistance or impedance is equal to the output impedance of the amplifier.

Practical conductors

It is usual to use copper wire, either insulated with a protective sleeving or not according to the circumstances, to connect together components in an electronic circuit. The wire is usually supplied in reels or coils and can be obtained in various gauges. The usual way to specify the diameter of the wire is according to the standard wire gauge (s.w.g.). For most circuitry work 20 s.w.g. wire has an appropriate thickness and pliability although where the wire is to be self-supporting a heavier gauge may be used, for example 16 s.w.g. Nearly all copper wire is tinned, that is covered with a coating of tin to assist in soldering the wire to the various components.

In the case of wire with an insulating sleeving to protect it from accidental contact with other parts of the circuit the sleeving is usually made of p.v.c., an insulating plastic material. The sleeving is available in different colours so that if a section of a circuit is wired in a particular colour it can be easily identified.

In much modern electronic equipment 'printed circuitry' is used instead of wire to connect the components together. The process of 'printing' circuits is described in Chapter 8.

Practical batteries

Different types of batteries are available for different requirements. Large secondary cells such as those used in motor cars have a capacity of about 80 ampere-hours, i.e. they can deliver up to 20 amperes for four hours without any appreciable change in their output voltage (or e.m.f.). This type of battery uses diluted sulphuric acid as its electrolyte, the probes taking the form of lead plates to provide a substantial surface area to the electrolyte—a factor which determines the electrical capacity of the battery. Each cell will provide an output voltage of 1·25 volts, this being determined by the electro-chemical nature of the cell. To obtain higher voltages a number of the cells can be connected in series, positive terminal to negative terminal and so on. If a number of the cells are connected in parallel, that is all the positive terminals together and all the negative ones together, the output voltage will remain the same as for one cell but the capacity will increase by a factor equal to the number of cells connected in this way. Secondary cells are very useful in laboratories where a large current at a constant voltage is needed.

Most of the smaller batteries are primary cells. With the use of transistors and semiconductor devices in modern equipment, ranging from medical instrumentation to guided missile telemetry, the use of small primary cells has greatly increased during the last few years. The electrolyte used is usually a paste of some kind, hence the term 'dry cell'. This type of cell can be physically very small and yet still provide a useful performance.

Standard cells are another important type of battery. They are designed to give a constant output voltage which should not change over a temperature range of $-30°$ C to $+70°$ C by more than $\pm\frac{1}{2}$ per cent, with a current drain on the battery of about 100 microamperes. These figures are obtained by designing a cell with a very low internal resistance. They are used by electronics engineers as reference standards when calibrating instruments, which may themselves be used later to make measurements on electronic and other types of laboratory apparatus.

Resistors

Resistors are made of a mixture of conducting and insulating material. In manufacture the proportions are adjusted to give resistances of various values.

Three types of composition resistor are commonly used: the lacquer type, which is mostly employed in variable resistor design; the resin-bonded type, the most common type of commercial resistor; and the ceramic-bonded type, an expensive type used where close tolerance is necessary.

Resistors are formed by taking the right proportions of conductor, insulating filler and bonding material and extruding this mixture to form a long cylinder. This is then cut to the required length, connecting wires are fastened to the ends, and it is finally vitrified in an oven.

The colour code and preferred values

To make resistors easily recognizable a system of colour coding is used.

A typical example is shown in Fig. 11, the colour bands being marked A, B, C and D. The colour code is easy to remember and

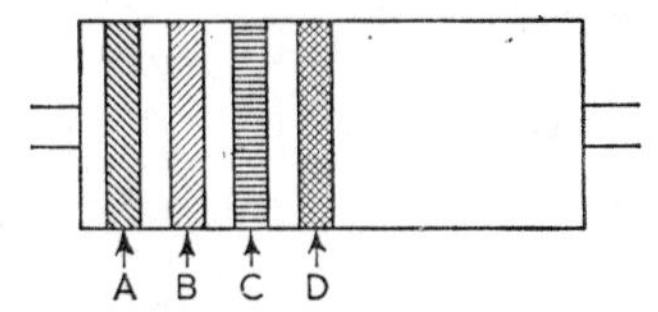

Fig. 11. The four band system of resistor colour coding.

with practice the reader should be able to give instantly the value of a resistor. The first colour band, A, indicates the first figure of the resistor's value, the second band, B, gives the second figure, and the third band, C, denotes the number of zeros which follow these first two figures. The fourth band, D, gives the tolerance rating of the resistor. The standard colours are listed in Table 2. As an example, if band A is red, band B violet and C red, the resistor's value will be 2 (red), 7 (violet) and two zeros (red) ohms, that is a 2,700 ohm resistor. With no fourth band the tolerance is 20 per cent: this means that the actual value could lay anywhere between 2,160 and 3,240 ohms. This would in practice be called a 2·7 k resistor (a kilohm equals one thousand ohms).

Because of this tolerance rating, commercial resistors are usually supplied in preferred values, for it would be pointless to have resistors whose tolerance bands overlapped one another. Typical preferred values are 120 ohms, 180 ohms, 270 ohms, 1·2 k, 1·8 k and 2·7 k, and so on, see Table 3. If the tolerance bands of these resistors are worked out it will be seen that they just touch one another. If it is necessary to have a resistor with a value other than the preferred ones, then it is necessary to take a selection of resistors of the nearest preferred values and measure each individually to find one of the required value. In practical circuit design such a procedure is seldom necessary, the nearest preferred value generally being good enough.

Table 2. Four-band Colour for Resistors

Colour	*1st Figure 'A'*	*2nd Figure 'B'*	*Multiplying Value 'C'*	*Tolerance 'D' (%)*
Silver	—	—	10^{-2}	±10
Gold	—	—	10^{-1}	± 5
Black	—	0	1	—
Brown	1	1	10	± 1
Red	2	2	10^2	± 2
Orange	3	3	10^3	—
Yellow	4	4	10^4	—
Green	5	5	10^5	—
Blue	6	6	10^6	—
Violet	7	7	10^7	—
Grey	8	8	10^8	—
White	9	9	10^9	—
None	—	—	—	±20

Table 3. Preferred Resistor Values and their Associated Tolerances

Tolerance			*Tolerance*		
±5%	±10%	±20%	±5%	±10%	±20%
1·0	1·0	1·0	3·3	3·3	3·3
1·1			3·6		
1·2	1·2		3·9	3·9	
1·3			4·3		
1·5	1·5	1·5	4·7	4·7	4·7
1·6			5·1		
1·8	1·8		5·6	5·6	
2·0			6·2		
2·2	2·2	2·2	6·8	6·8	6·8
2·4			7·5		
2·7	2·7		8·2		
3·0			9·1		

2

DIRECT AND ALTERNATING CURRENTS

In the last chapter we talked about an electric current and how it can be produced. This type of current is called direct current (d.c.) since it flows in one direction around a circuit. A graph of the current drawn against time, that is the value of the current at given points in time, might appear as shown in Fig. 1. A practical use of such a graph would be in the recording of changes in amplitude of a current over a period of time due to some variation in the e.m.f. source or in the resistive behaviour of the load. But whatever the purpose, the graph shows that at intervals of time (taken on the stopwatch) the reading on the meter (indicating the current) alters by a slight amount. The graph is therefore the *waveform* of a direct supply current (or voltage).

This term, waveform, is one commonly used in electronics. Every current or voltage changes to some extent with time and thus has a waveform which in many cases is repeated over and over again.

The most common waveform is the sine waveform. This is a well-known mathematical relationship—it goes on repeating itself over a period of time with a shape as shown in Fig. 2.

The waveform of any energy source may be distorted by the circuitry to which it is applied and hence it is common to hear engineers say that a waveform has been *distorted*.

Alternating current

The current obtained from the mains socket of an ordinary domestic power point is usually alternating current. This type of current is provided by motor generators which convert mechanical energy into electrical energy. The output of these generators is positive in one direction for a short period, then reduces to zero; the direction then quickly reverses and the current gradually increases again in the opposite direction. This complete change of direction is

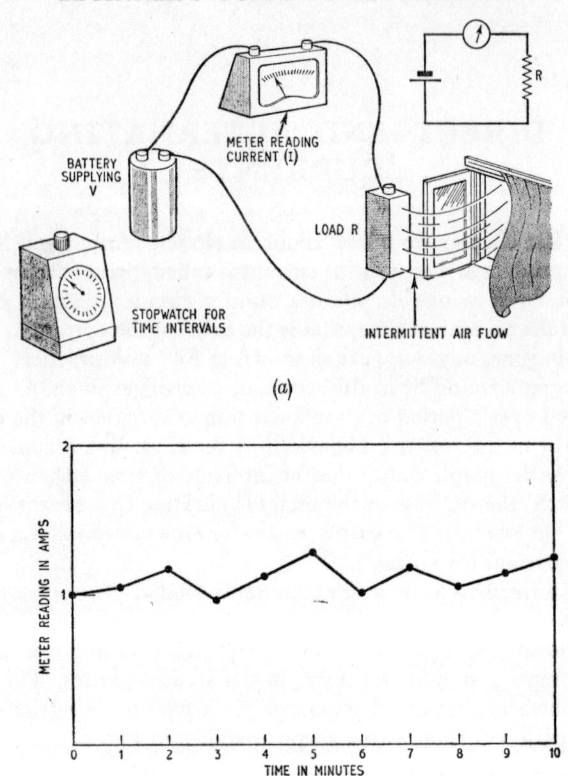

Fig. 1. Plotting a direct current over a period of time.

called a cycle and is shown in Fig. 2. These cyclic changes or waveforms occur rapidly, the standard for mains purposes in Great Britain being fifty per second. Each complete cycle is a sine wave. Cyclic motion occurs in many natural phenomena—a common example is the children's swing which as it moves back and forth with a varying amplitude describes what is called simple harmonic motion. If the distance moved by the swing is plotted against time a sine wave like the one shown in Fig. 2 (b) would result.

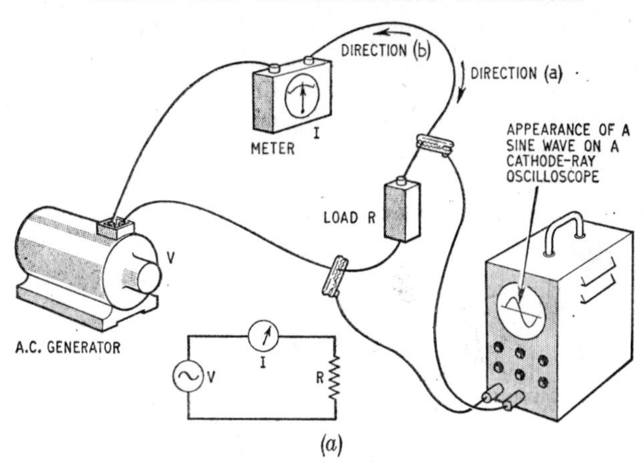

(a)

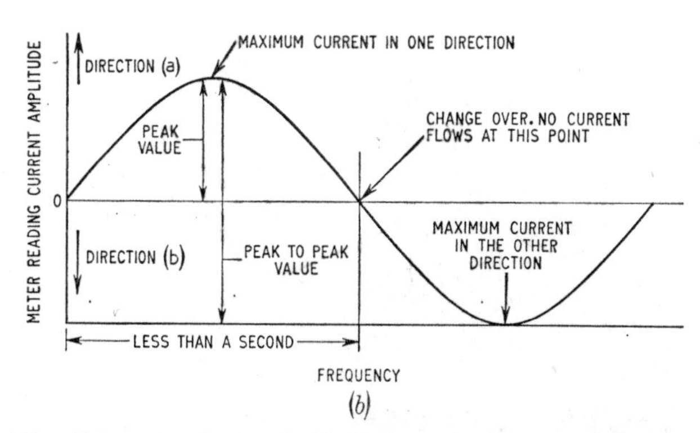

(b)

Fig. 2. One complete cycle of alternating current. The waveform (b) is sinusoidal.

This sinusoidal current has a number of advantages over direct current, the most important one being that obtained by using a transformer. Currents and voltages may be changed in value and relationship with comparative ease by this means. In this way mains power can be distributed at a high voltage and low current so that the heat which would be produced by a large current flowing in the copper conductors is minimized. When the supply cable arrives near the domestic area the voltage is reduced to a safe level and current can be drawn by the user without causing great losses. This can only be done by using alternating current and transformers.

In radio communications it is necessary for radio energy to be emitted by the transmitting aerial and radiated into the ether. This is done by using alternating currents with cycle times so small that

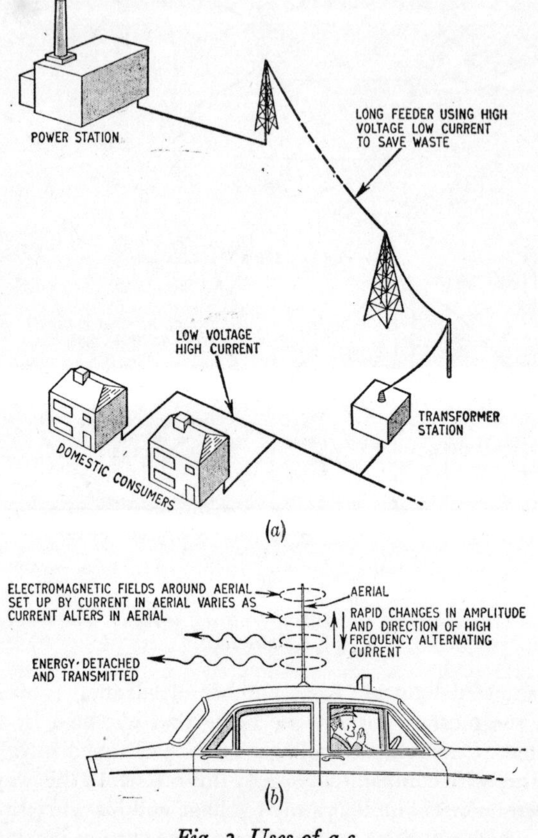

Fig. 3. Uses of a.c.

they may occupy less than many thousandths of a second. These rapid changes of current in the aerial generate electromagnetic fields which become detached from the aerial carrying the current. In this way the radio energy is emitted and radiated.

These two examples are illustrated in Fig. 3.

There are several important parameters to be considered in regard to sinusoidal alternating current (a.c.). These are amplitude, frequency and phase.

Amplitude

Amplitude may be referred to in a number of ways. Referring again to Fig. 2, it is possible to talk about the peak amplitude of the cycle (which may be either current in amperes or e.m.f. in volts) or the peak to peak value. The latter is often used in television practice, where it is called the DPA, double peak amplitude. The third and most common way to refer to the amplitude of a sine wave is to talk of its root mean square value (r.m.s.). This value is 0·7 times the peak value and is equal to the amplitude of a direct current which, flowing for the same time in the same conductor with the same resistance, would generate the same amount of heat.

Frequency

In electronics the frequency of alternating currents is an important quantity. The term is used to describe the number of complete cycles that occur in a given period of time—usually one second. The alternating current distributed for domestic use completes one cycle fifty times in a second. This is usually written as 50 c/s. The current alternating in an aerial may complete the cycle in less than a microsecond (a millionth of a second) and therefore its frequency may be several megacycles (millions of cycles) per second. This is usually written as Mc/s.

The electronics engineer generally deals with current changes which vary from 50 c/s up to 10,000 Mc/s or higher. This latter frequency is so high that it becomes hard to think about, or to express, in analogous terms.

The term frequency occurs often and it is the behaviour of circuitry at various frequencies that concerns the engineer. Some

Table 1. Frequency bands

Description	*Frequency*	*Wavelength*	*Application*
Audio frequency (A.F.) (also very low frequency V.L.F.)	10–30,000 c/s	—	Sound reproduction
Radio frequency (R.F.):			
Low frequency (L.F.)	30 kc/s to 300 kc/s	10,000–1,000 metres	Communications, R.F. heating
Medium frequency (M.F.)	300 kc/s to 3 Mc/s	1,000–100 metres	Telemetry, control, communications
High frequency (H.F.)	3 Mc/s to 30 Mc/s	100–10 metres	Communications
Very high frequency (V.H.F.)	30 Mc/s to 300 Mc/s	10 to 1 metres	Communications and control
Ultra high frequency (U.H.F.)	300 Mc/s to 3,000 Mc/s	1 to 0·1 metres	Television, communications, navigation
Super high frequency (S.H.F.)	3,000 Mc/s to 30,000 Mc/s	10 centimetres to 1 centimetre	Radar, special work

examples of this are the transmission of many frequencies simultaneously over cables and the difficulty of operating valves and transistors at very high frequencies. In some spheres of electronics, for example in radio communications, it is the problem of handling these high frequencies that bothers the engineer, whereas in medical electronics, for example, it is the problem of handling very low frequencies that often causes difficulties.

In Table 1 the terms used for the main bands of frequencies are given. It should be noted that the term radio frequency (r.f.) which is often used is a loose term applied to electric energy alternating at any frequency above about 10 kc/s.

Wavelength

In Fig. 4 (a) a sine wave, that is a complete cycle of alternating current, is shown to demonstrate the relationship between fre-

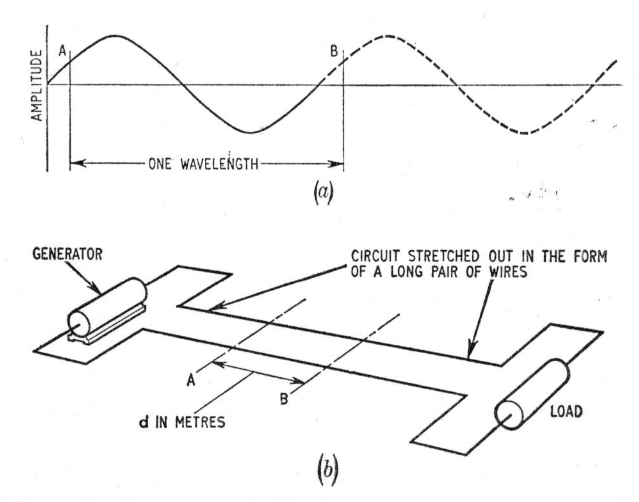

Fig. 4. Wavelength. (a) Two successive cycles. (b) Wavelength on a pair of wires.

quency and wavelength. So far we have discussed a current alternating in a complete circuit formed by the generator, the load and the wires connecting them. In this type of circuit it is easy to understand a current flowing first in one direction and then decreasing to zero and flowing in the other direction. The electrons are being moved back and forth in the manner of shingle lying on the beach under the influence of the tide. However, if the circuit is made physically big enough (or the frequency made high enough) the electrons are still executing one movement when the e.m.f. changes direction. Odd effects result and the wavelength becomes an important unit in which to measure these effects.

The wavelength of a sine wave is the distance (usually measured in metres) between two points on exactly similar parts of successive cycles, for example points A and B in Fig. 4 (a). A practical example of wavelength occurs if a circuit is physically stretched out in the form of a long parallel pair of wires, as shown in Fig 4 (b), with a generator of the sine waves at one end and the load at the other. The voltage will be present along the wires (if they are long enough, or, if the frequency is high enough) as a series of maxima and minima. The distance between successive maxima or successive minima corresponds to a wavelength. This distance is related to frequency by the velocity of the electrical energy along the wires. Since under normal conditions the velocity is constant, the higher the frequency the smaller the wavelength. When dealing with the electrical energy in the U.H.F. and S.H.F. bands (see Table 1) it is usual to speak of wavelength and not frequency. The engineer talks about 3 centimetre waves and not 10,000 Mc/s frequency. The problems encountered at these frequencies are the concern of the microwave engineer who deals with U.H.F. and S.H.F. energy.

Phase

In Fig. 5 the relationship in time between an applied e.m.f. and the resultant current is shown for a sine wave. Since they are displaced they are said to be out-of-phase (the current that flows is displaced in time from the applied voltage). Each cycle is divided into 360° (360° is also equivalent to the distance of a wavelength) and the lag or lead of the current relative to the applied voltage is measured in degrees. In Fig. 5 it is 180°. Many factors can cause an alternating voltage and its current to be displaced.

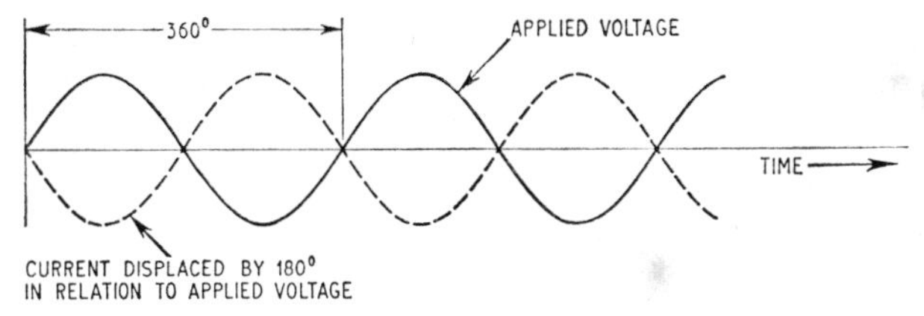

Fig. 5. Phase relationship between current and voltage.

The importance of phase

We can get a clearer idea of the importance of electrical phase if we consider a mechanical analogy to it. A large, heavy handcart is difficult to move: the force needed to move it has to be applied before any movement takes place and is greater than the force required to keep it in motion. This is a case of the *effect* lagging behind the *cause*. If, on the other hand, a water pump is connected to an empty tank water will pour freely into the empty tank even before the pump applies any force and it is not until the tank begins to fill up and produce 'back pressure' that the pump has to work to continue the flow of water to fill the remainder of the tank. In this case the effect leads the cause.

In an electrical circuit the applied voltage is the cause and the current that flows is the effect, since one cannot have current flowing without a voltage being applied, although one may have a voltage applied without any current flowing.

In an electrical circuit the current can lead the voltage applied to the circuit or it can lag the voltage. If it leads it is usually because the source of voltage is applied to a capacitor (see later) which, like the empty water tank, lets the effect take place before the cause. If the current lags it may be because the voltage is applied to an inductor (see later) which is like the heavy cart in its behaviour: this time the voltage has to be applied for some definite time before the current flows.

In both cases the voltage and current do not act together and are said to be *out-of-phase* with each other. The amount by which they differ (which can be as much as a quarter of a cycle in the case of a sinusoidal wave) is referred to as the *phase difference*.

Pulse waveforms

So far we have discussed two types of current, direct and alternating. A third type has some of the characteristics of both and is called pulse-waveform current.

In its simplest form it can be produced by the circuit shown in Fig. 6 in which a battery is switched on and off continuously. A

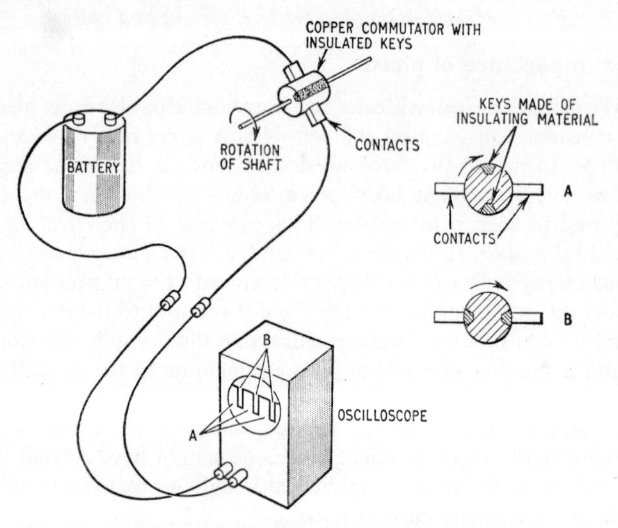

Fig. 6. Simple method of producing a pulse waveform by means of a rotary switch (commutator).

graph can be produced by measuring this current (either with a chart recorder or an oscilloscope) and plotting it against time. Fig. 7 shows this.

Looking at this graph one can see certain features of this current which are important: (1) It is direct current, that is it flows only in one direction. (2) It is intermittent, and thus shares to some extent a characteristic of alternating current. (In particular any current that changes, whether smoothly as in a.c. or rapidly as in pulsed current, can be used to transfer energy through a transformer.)

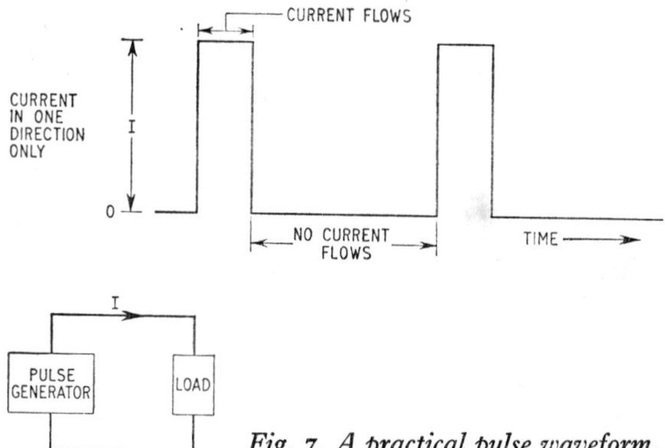

Fig. 7. A practical pulse waveform.

Describing the pulse

To describe a pulse-waveform we do not talk about r.m.s. value as in a.c. but usually discuss the *peak value* as shown in Fig. 8 (a).

Whereas with a.c. we discuss frequency, when dealing with pulses it is usual to state the number of pulses per second and this is called the *pulse repetition frequency* (p.r.f.). Fig 8 (b) demonstrates this for a p.r.f. of 100.

Each pulse has to be a certain length and there are intervals between pulses. The pulse length is called the 'mark' time and the intervals between pulses are termed the 'space' time. Hence the *mark/space ratio* is an important parameter when discussing the shape of pulses (Fig. 8 (c)).

A pulse must begin at some point and then return to it for a space interval. This point is called the *reference level* and varies in different circuits. In fact the reference level of a pulse can be altered by passing the pulse through a suitable circuit. The problem of maintaining or deliberately altering a reference level is one of considerable importance to the electronics engineer.

Although we have talked about pulses of current, it is often necessary to deal with a voltage pulse which causes virtually no current to flow. This may be applied to the input of an amplifier or counting device. Some current does flow under these conditions,

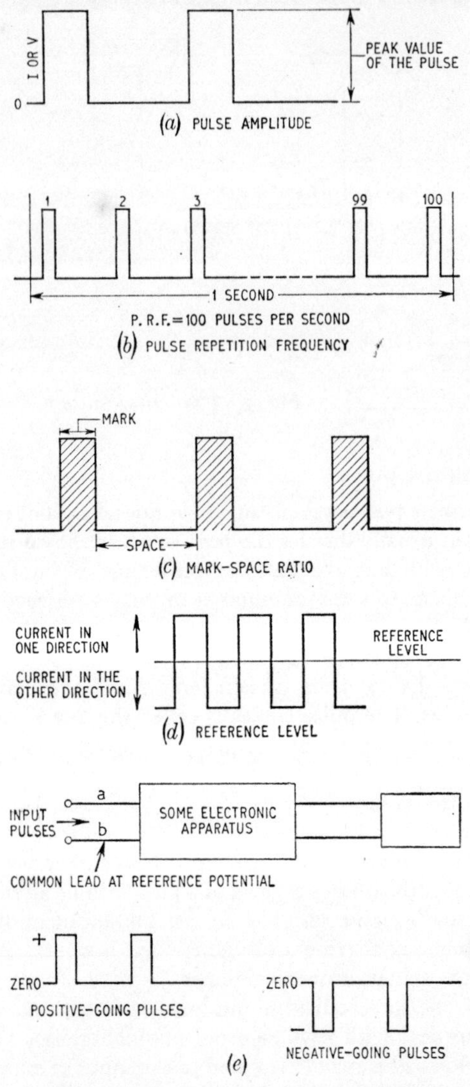

Fig. 8. Pulse waveform characteristics.

but is so small—for example a few microamperes—that the engineer tends to disregard it.

To describe the action of these applied pulses in any system it is necessary to know their *polarity*. As a practical example of the importance of this a circuit designed to operate with pulses of one polarity may be able to reject a pulse of the opposite polarity. The two polarities are distinguished by calling pulses *negative-going* or *positive-going*. It can be seen from Fig. 8 (e) that if one terminal of the input to an apparatus is connected to the chassis of the equipment to which the pulse is applied and this is kept at zero potential to earth (that is no volts above earth) then this terminal (b) is at the zero reference level and when no pulses are present, that is during a space, the other terminal (a) is also at the zero potential. If a pulse comes along and makes terminal (a) positive with respect to (b) then the pulse is said to be positive-going, whilst if it makes (a) negative with respect to (b) it is called a negative-going pulse.

In discussing circuit design an engineer may refer to a negative-going squarewave with an equal mark/space ratio and a p.r.f. of 500 pulses per second—this being a complete description of a particular pulse waveform.

Pulse distortion

A certain amount of distortion to a pulse when it is passed through an electronic circuit is inevitable. This is caused by various factors in the circuit which delay the pulse. It is an important part of the work of the designer today to devise means of manipulating delays or time lags. Four terms are in common use to describe these delays, *delay time*, *rise time*, *storage time and fall time*. Fig. 9 shows at (a) a typical pulse and at (b) its shape as an output pulse after being passed through an electronic circuit. Delay time is the initial time taken for the output pulse to reach 10 per cent of its final maximum value. This is usually a very small time, for example 0·3 microseconds (μsec). More important are the terms rise time and its converse fall time, which are the times taken for the output pulse to rise and fall from 10 to 90 per cent of its total value. By ignoring the 10 per cent 'tail' portions of the pulse the engineer concerns himself only with the major change that takes place to the pulse.

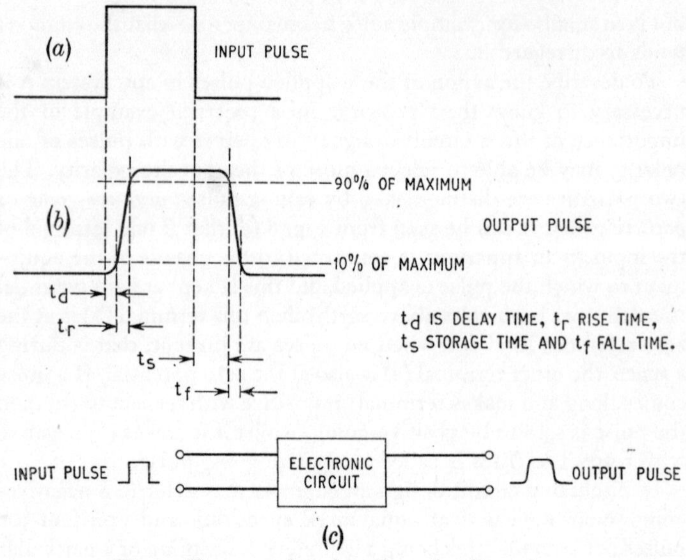

Fig. 9. Terms used to describe pulse distortion.

Storage time is the time delay caused by the circuit not cutting off immediately upon the cessation of the input pulse. Typical rise and fall times are 1 μsec, a typical storage time being 4·5 μsec., these times applying to the reaction of a single fast-acting transistor to the presence of a pulse.

The sine wave and the pulse

Every pulse waveform can be analysed into a number of sine waves or, conversely, a number of sine waves can be synthesized to make a series of pulses as shown, although in a rather over-simplified way, in Fig. 10. This makes things a bit easier because all the rules for sine waves will also apply to pulses.

Since a number of sine waves will combine to produce a pulse with a certain repetition frequency it is obvious that one of these frequencies must dominate and determine the repetition rate. This frequency is called the *fundamental frequency*.

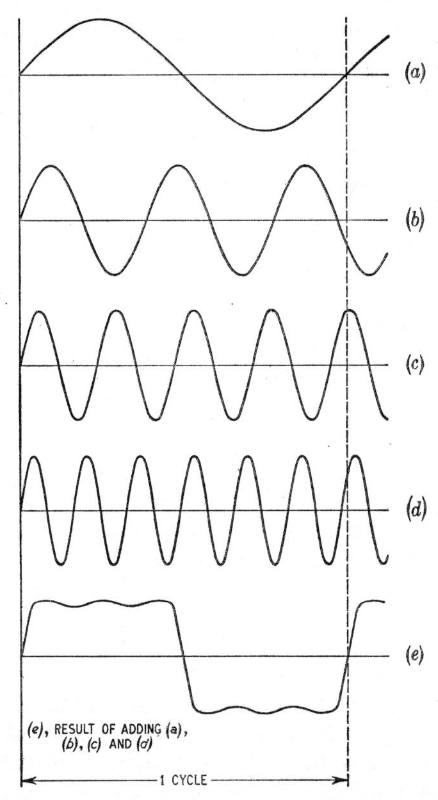

Fig. 10. Pulse waveform (e) derived by adding up a number of sine waves of different frequency.

Other types of waveform: the sawtooth and the spike

There are other types of waveform which the reader may encounter. Amongst them the sawtooth is perhaps the most important. This is shown in Fig. 11 (a). The waveform gradually increases in value as time increases: after reaching a maximum value it then falls in a very short period of time (called *flyback time*) to the zero line or reference level. The cycle is then repeated.

Another waveform commonly encountered is the spike waveform shown in Fig. 11 (b). This is really a square pulse with a small mark/space ratio. When using a spike waveform the electronics engineer is only concerned with the leading edge and does not mind

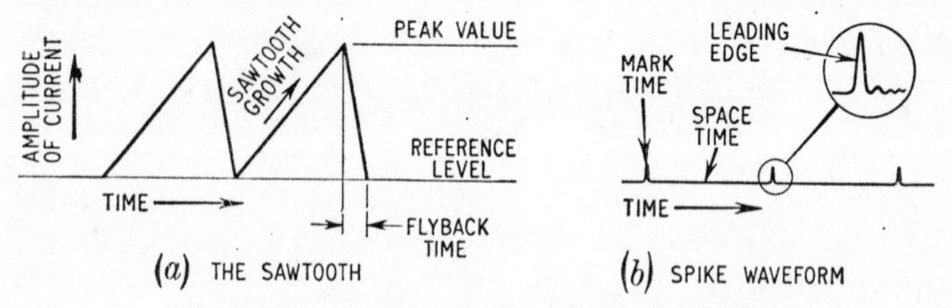

Fig. 11. Two other common types of waveform.

any distortions (such as long-tails) that may occur during the long space interval.

Alternating and pulse voltages, whether square, sawtooth, spike, etc. will all be applied to circuits and as a result of their application current will flow. How much current will flow is no longer determined by resistance alone but other factors to which we must now turn.

The effects of a changing voltage when applied to a circuit

Just as in the case of direct currents, the amplitude of alternating and pulse currents is determined by the amplitude of the applied e.m.f. and the resistance of the conductor. However there is an additional factor which opposes the flow of alternating or pulse current and it is the result of the alternating character of such current. It is called *reactance* and its value is determined by the characteristics of the circuit through which the current flows. Every conductor possesses, to some extent, two electrical characteristics called capacitance and inductance. These characteristics can be varied by making the conductors take up certain physical shapes, e.g. by winding a wire into a coil (making an inductor) or by placing flattened conductors near to one another (the plates of a capacitor).

Both these characteristics have the effect of limiting the amplitude of the current flowing in a circuit when an alternating or pulsed voltage is applied. This effect is called reactance and its magnitude, which is measured in ohms (in exactly the same way as resistance), is determined by the amount of inductance or capacitance present in the circuit.

The value of reactance is also dependent on the frequency of the a.c.—the higher the frequency the higher is the inductive reactance and the lower the capacitive reactance. In certain configurations these reactances oppose one another and cancel out either the whole or part of each other. Reactance combines with the resistance to give *impedance.* So when dealing with alternating currents we are concerned with impedance and not just resistance.

Inductance, capacitance and resistance values depend on the physical shape of conductors, but their electrical effect, which causes impedance to current flow, does not come into play unless the current is varying. With direct currents (which are zero frequency) only resistive effects are normally operative.

Inductance

When a current flows in a length of conductor it produces a magnetic field around the conductor as shown in Fig. 12. The

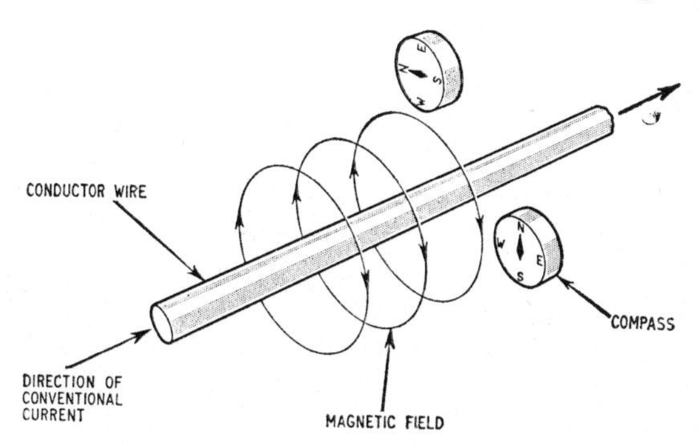

Fig. 12. Magnetic field round a conductor carrying a current.

influence of this field can be easily detected: a compass needle held near a wire carrying a heavy current will be deflected by the field the current sets up. If the current carrying conductor is placed adjacent to another separate conductor, as shown in Fig. 13, and the current is increased in amplitude, then the field surrounding the first conductor will 'swell' out and 'cut through' the second wire. In doing so it will induce an e.m.f. and hence a current in the second wire conductor.

The direction of the induced e.m.f. and its current will be such as to produce a second field. This second field will couple with the first wire to generate a third e.m.f. This 'reintroduced' e.m.f. is in a direction that will oppose the original current in the first wire. It is almost as though the second conductor was saying 'please leave me

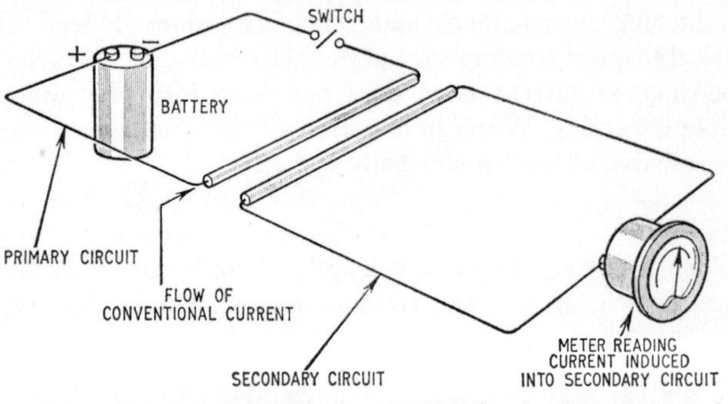

Fig. 13. Mutual inductance. The current flowing in the primary circuit causes a current to flow in the adjoining secondary circuit.

alone'. Its behaviour is one of reluctance to be disturbed. This is known as an inductive effect and the two conductors in this position are said to have the property of *mutual inductance.*

This effect of inductance is an important one in electronics. Even a straight piece of conductor has some self-inductance though it is not wound in a coil or placed near another conductor. The amount is very small, but as the frequency at which the current alternates is increased the effect of even this small inductance becomes more important.

The opposition to the flow of an alternating current through an inductor increases as the frequency of the alternating current increases. The combined effect of inductance and frequency is called *inductive reactance* and is measured in ohms. It cannot be added directly to resistive ohms—the combined effect of the two when placed in series is slightly less than the arithmetic sum of both.

It will be seen that a coil carrying a direct current of constant amplitude might just as well be a straight piece of wire but, as soon as the current begins to vary, as is the case with alternating current or pulse waveforms, then the inductance begins to show itself as inductive reactance.

The effect of inductive reactance on a pulsed waveform is a little more complicated. Since this type of waveform can be considered as a large number of sine waves of different combinations, it is clear that the inductor will present different reactances to the different frequencies and limit some more than others. The result is that a pulsed waveform current may be distorted in shape after passing through an inductor.

Units of inductance

Inductance is measured in a unit called a Henry. In electronics one deals mainly in millihenries (thousandths of a henry, abbreviated mH) or microhenries (millionths of a henry, abbreviated μH). The symbol used in circuitry to denote an inductance is L.

Uses of inductors in electronics

The property of inductance is used a great deal in electronic circuits. Often it is necessary to take steps to minimize its effect, as in the case of large radio transmitting valves which when operating at high frequencies can distort the signal they are sending out if inductance is present in the wrong place. A common use of inductors is in radio receivers where, combined with capacitors, they are used to select the frequency it is desired to receive from the many available. The ability of inductors to change their reactance according to frequency means they can be used as filters to block off one frequency and allow another to pass.

Inductors are usually coils of wire wound on some insulated support called a coil former. This type of coil is suitable for many purposes but in some instances it is desirable to have a relatively high inductance with a few turns. This can be achieved by concentrating the magnetic field through the coil by placing a magnetic core in the former as shown in Fig. 14 (a).

When an inductor is used at very high frequencies, unwanted currents induced in the magnetic core may reduce the 'goodness'

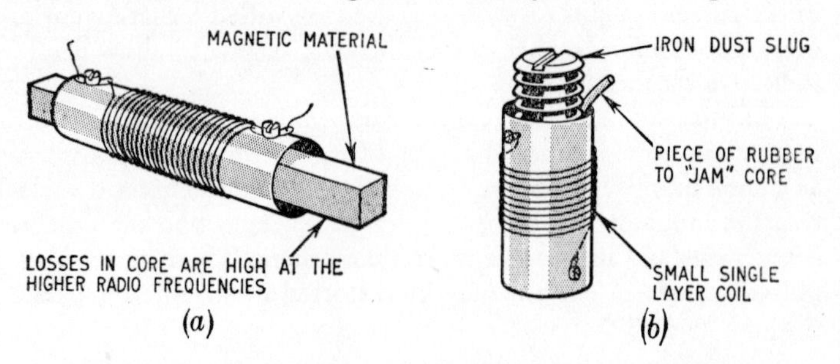

Fig. 14. (a) Inductor with core. (b) Typical dust core inductor.

factor, or *Q* of the inductor. To avoid this the following technique is used: the magnetic material is crushed into fine particles and mixed with an insulating powder and binding agent. The resultant iron dust-cores will concentrate the magnetic field but prevent unwanted currents flowing. A typical dust core and coil is shown in Fig. 14 (b).

Q factor

Fig. 15 shows the equivalent circuit of an inductor. As can be seen there is some resistance in series with it. This resistance is that of the wire itself, and will be effectively increased by losses caused by unwanted currents flowing in the core and in any screening that may be placed around the coil. In any circuit if energy is lost in some way it is usual to regard this as being due to an imaginary resistor or, as it is called, an *equivalent resistance* in the circuit. A perfect coil would offer no resistance, having reactance only. Practical coils however offer both, for the reasons just given. The ratio of the reactance to the resistance of a coil is a measure of its 'goodness', and is termed the *Q* factor of the coil. The aim in designing inductors is to keep the ratio of the reactance to the

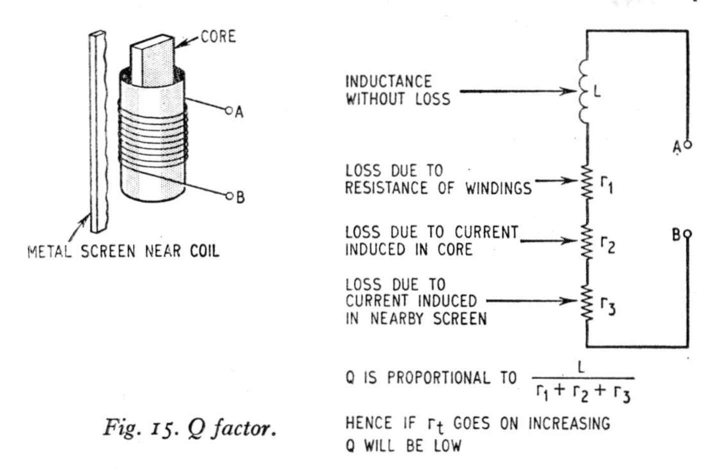

Fig. 15. Q factor.

resistance as high as possible. In a perfect coil, Q would be equal to infinity but in practice a coil designer may be satisfied with a Q factor of several hundreds.

Practical inductors

Fig. 16 shows some of the types of inductors used in electronics. The first (a) is a typical one for radio transmitters. Its value may only be millihenries or microhenries since it will be used for frequency selection or tuning purposes at ultra high frequencies. The wire in this instance is a silver plated tube since at ultra high frequencies the current flow in a conductor tends to be concentrated in its 'skin' (hence the term *skin effect*). Each turn is carefully spaced from the other so that although the magnetic fields between turns will couple there will not be any unwanted capacitance (see later) between adjacent turns. The coil former is usually made of a ceramic material to avoid losses and so produce an inductor with a high Q.

The second inductor (b) has an iron core made of laminated iron, and a large number of turns. Its Q is very low but this does not matter for the purpose for which it is intended. It is used as a filter in power supplies and for similar applications.

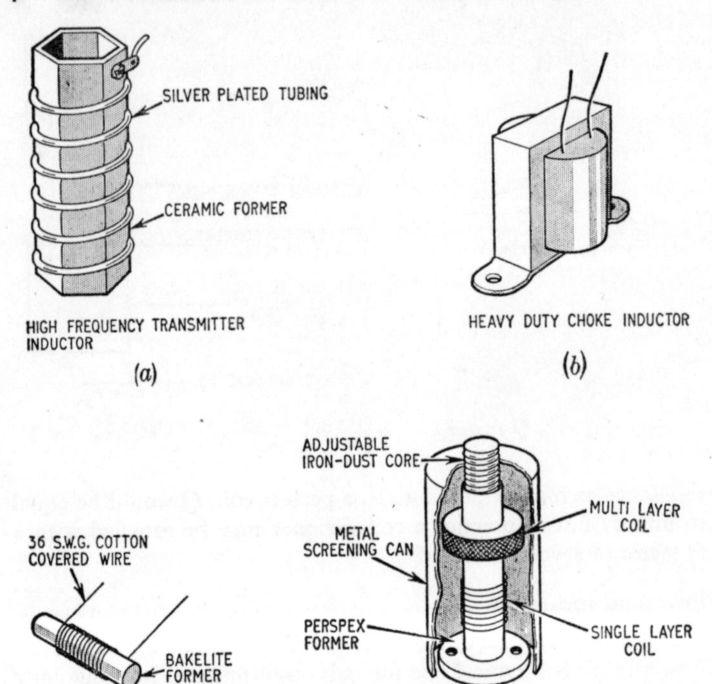

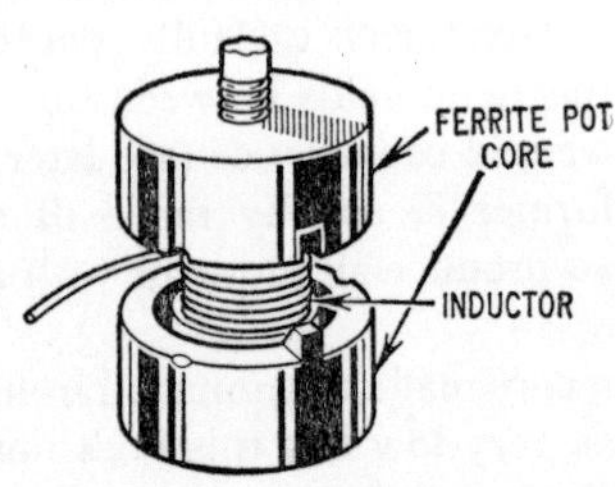

Fig. 16. Common types of inductor.

Inductor (c) is a simple filter coil used to block-off radio frequency currents in some part of a circuit. Although it is shown on a former often an engineer will wind one using a high value resistor as a former, this combination being most suitable as a small r.f. choke.

Fig. 16 (d) shows a tuning coil in a screening can with an adjustable dust core. This type may be found in radio communication equipment, television and many other applications. The remaining inductor (e) is one of the latest and is used in many filter circuits, pulse forming networks and so on. The small coil is totally enclosed in a ferrite pot which can be screwed up tightly. This type of inductor gives a high inductance—many hundreds of millihenries (a high value in electronics) for only a few turns of wire, thus providing a physically small coil with a high inductance and high Q.

Capacitance

Like inductance, capacitance only becomes manifest when the amplitude and/or direction of the current in a circuit is changing.

The component called a capacitor may consist of two plates near to one another but separated by an insulator such as air. This arrangement is shown in Fig. 17 (a), the two plates being shown connected to a battery.

When the plates A and B (made of conductive material) are first connected to the battery electrons are drawn from the plate A towards the positive terminal of the battery. This leaves a deficit of electrons on this plate which causes an electric field to exist in the insulator, the air, D, separating the plates. This electric field will cause electrons in plate B to migrate inside it and thus the outside of this plate (B) will attract electrons from the negative terminal of the battery, as shown in Fig. 18 (b). This movement of electrons is a current and so it seems as if a current is flowing in a circuit which is broken by a section of insulating material. This is in fact true, and if the plates were infinite in area and the gap D extremely small (provided the e.m.f. applied did not cause an actual breakdown, i.e. spark, across the gap), then current would flow for as long as the battery lasted in exactly the same way as it would in a normally closed circuit. Such an arrangement would be said to have an infinite

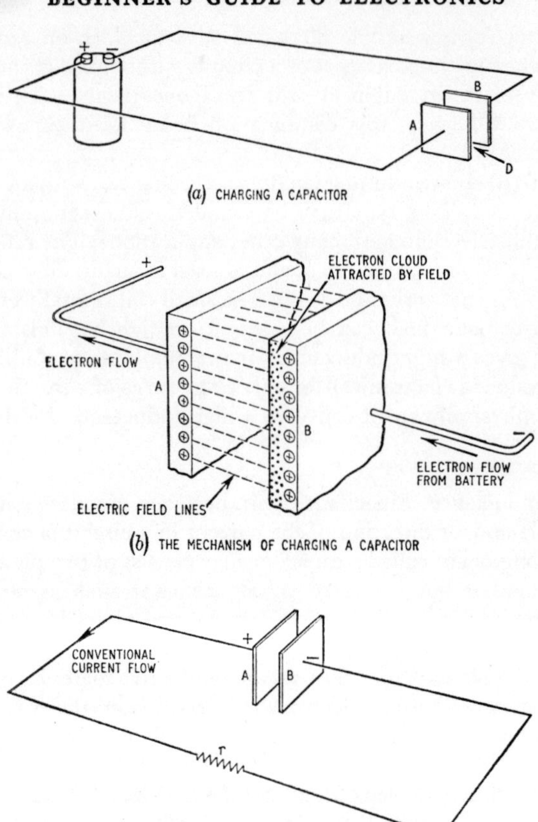

Fig. 17. Effect of capacitance.

capacity. However, in practice the capacitor's plates must be of finite size and this means that the current will flow only until the charge on the plates produces a voltage across the plates equal to the e.m.f. from the battery. When this point is reached no more current will flow.

The current that flows into a capacitor, infinite or finite, is stored as an electric charge which is measured in coulombs (see Chapter 1). If at some point in time the battery is disconnected it will be found that the capacitor now acts like a secondary cell in that it will return these coulombs of electricity as a current if the two plates are joined by a conductor as shown in Fig. 17 (c). The e.m.f. at which it starts to do this (i.e. discharge) is equal to the e.m.f. which originally was used to charge it up but decreases gradually (exponentially) as the number of coulombs stored runs out.

The charge stored in any capacitor is determined by the physical size of the capacitor, the applied e.m.f. (and hence the current flowing) and the time for which the current flows. The larger any of these factors are, the greater the charge stored.

The value of a capacitor (its capacitance) is measured in Farads. This unit is very large and in practice the electronics engineer usually deals in microfarads (millionths of a farad, abbreviation μF, or picofarads (millionths of a microfarad, abbreviation pF). The symbol used in diagrams for capacitance is the capital letter C.

Capacitance and alternating currents

If instead of disconnecting the capacitor from the battery and joining the plates with a piece of copper wire we reverse the battery terminals, as shown in Fig. 18 (a), we get a situation in which the capacitor will find that the existing charge it has on its plates actually aids in the process of recharging. If the battery voltage is altered in a sinusoidal manner, that is, the first e.m.f. is reduced to zero before the polarity is reversed, then the discharge current belonging to the first e.m.f. forms the charging current for the second e.m.f. This is explained in Fig. 18 (b).

To an alternating e.m.f. the capacitor appears to conduct electricity but the current always leads the voltage. This can be illustrated by means of the analogy shown in Fig. 18 (c). If a pump is pumping water first in one direction in a pipe circuit and then in another direction—that is in an alternating manner—a diaphragm placed in the pipe circuit would not stop the apparent flow of water 'current'. As the pump force (e.m.f.) drives the water back and forth the diaphragm distends and for a moment stores some water which it returns to the circuit when the pump force direction changes.

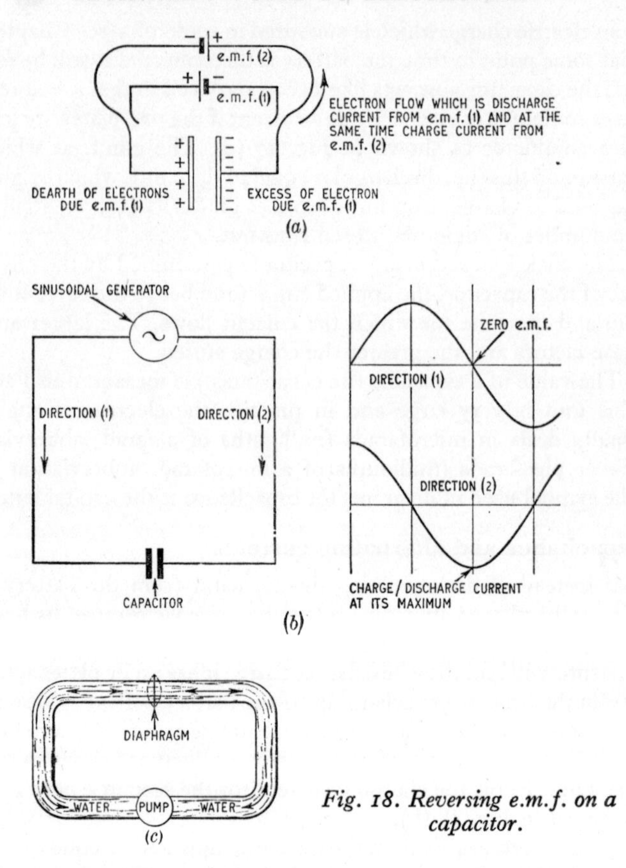

Fig. 18. Reversing e.m.f. on a capacitor.

This is exactly the way in which a capacitor acts in an a.c. circuit when it appears to be conducting the current in a circuit. The degree to which this conduction occurs is a measure of the *capacitor's reactance* and again is measured in ohms like resistance and inductive reactance. Although measured in ohms, like inductive reactance it cannot be added directly to resistive ohms. The higher the frequency and the larger the capacitance the lower the reactance.

Capacitance and pulse waveforms

Since a pulse can be analysed into a number of sine waves, pulses can be conducted by capacitors although the reactance will be different for the various sinewave components of the pulse. This means that the relative amplitude of the sinewave components is changed after being passed through a capacitor so that some distortion of the waveform occurs.

Practical capacitors

To explain the action of a capacitor we have assumed it to consist of two plates separated by air. In practice, however, it may be a more complicated device.

A simple air dielectric capacitor is shown in Fig. 19 (a). In this type the value of capacitance can be altered by varying the amount of plate enmeshed. This type of capacitor may have a capacitance only about 200 pF and is used for 'trimming' tuning circuits.

A larger but similar type, using many plates in parallel to increase the maximum capacitance value, can be found in the ordinary radio receiver.

A common type of capacitor (see Fig. 19 (b)) uses metal foil for its plates and paper as its dielectric (insulator). It is wound in a spiral which makes a small, tightly packed unit with a fairly high capacitance. This type can be obtained up to 100 μF or more, this being a high value in electronics.

A type of capacitor with special properties is the electrolytic. Although this capacitor breaks down easily if the applied voltage is too high, and can only work with one polarity of e.m.f. applied, it can be very useful to the engineer. Since the applied e.m.f. can not be reversed the electrolytic capacitor is limited to jobs where it smooths out ripples on top of direct current.

An electrolytic capacitor is shown in Fig. 19 (c), its case being cut away to show the interior. Two aluminium plates are immersed in a suitable electrolyte. The application of a direct voltage causes a thin insulating film of aluminium oxide to form on the positive electrode. This insulation becomes the dielectric and thus a capacitance is formed. Electrolytics are often used in preference to other

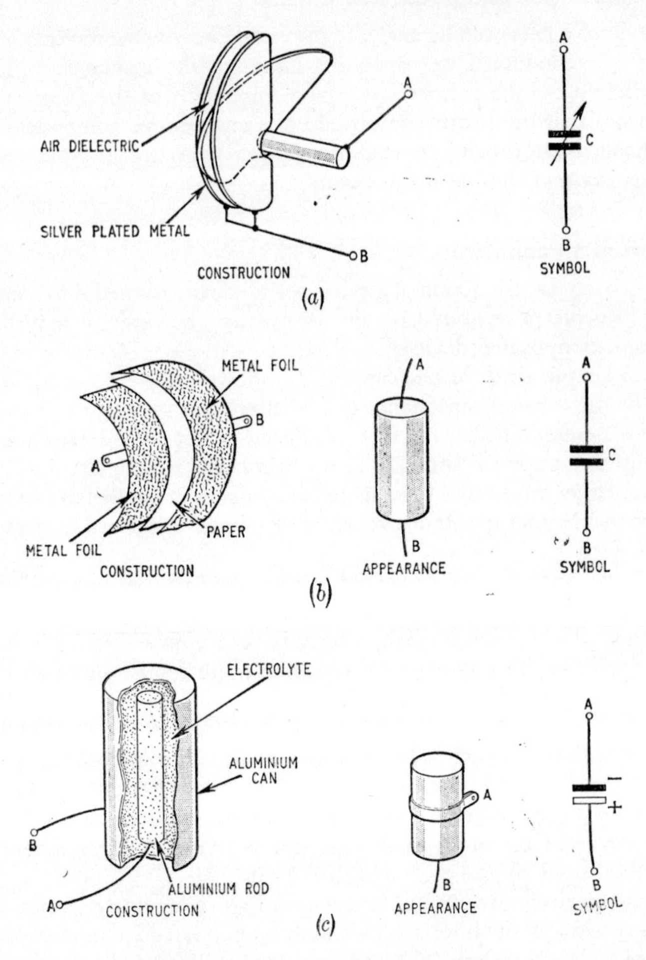

Fig. 19. Common types of capacitor.

types of capacitor because size for size it is possible to obtain a higher capacitance in an electrolytic capacitor than with other types.

Inductance and capacitance combined

Fig. 20 shows two combinations of *resonance circuit* formed from inductors and capacitors—series and parallel. Both may be used in electronics for frequency selection.

At any frequency other than the resonant frequency the first combination offers some impedance to the flow of current but, at resonance, the inductive reactance and the capacitive reactance

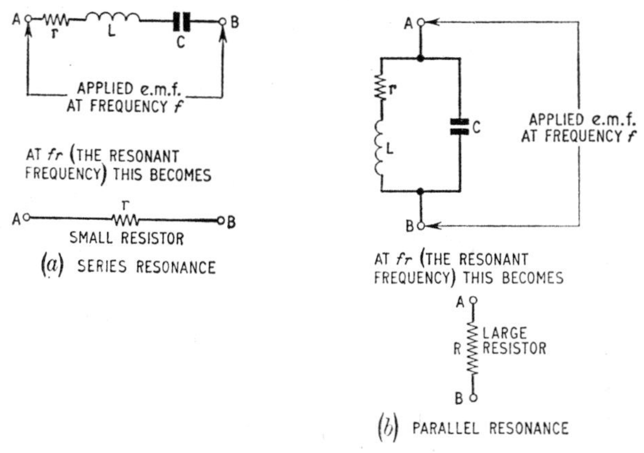

Fig. 20. The combination of inductance and capacitance produces resonant "tuned" circuits.

cancel out and the only opposition to current flow is the small resistance due mainly to the inductor windings (and maybe skin effect if the frequency of operation is very high). The inductive and capacitive reactances will cancel out at one frequency only, the one at which both reactances are numerically equal. This combination can therefore be used to allow a current at a certain selected frequency only to pass.

The second combination shows an inductor and a capacitor connected in parallel. At resonance the two reactances cancel out but this time the effect is that of a large equivalent resistance whose value is only made lower by the losses in the circuit due to the resistance in the inductor and perhaps skin effect. At all other frequencies this combination has a fairly low impedance. Thus this arrangement can be used to block current at a certain frequency.

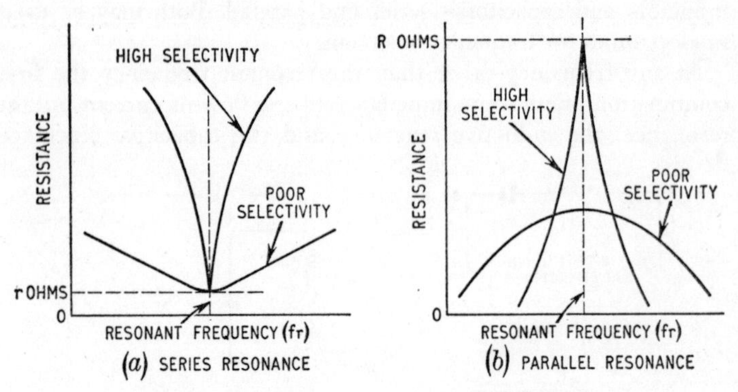

Fig. 21. Resonance curves of tuned circuits.

The sharpness of the tuning curves shown in Fig. 21 (for both series and parallel combinations) is called the *selectivity* of the circuit, i.e. its ability to select one frequency from amongst many. How well it does this is determined by the Q of the circuit, due to the Q of the inductor which we explained earlier.

3
ELECTRONIC COMPONENTS

Every circuit consists of an arrangement of components. Some of these—resistors, capacitors and inductors—we have already discussed. Amongst the many other components to be found in electronic circuits are thermionic valves, semiconductor devices, relays, neon tubes, counting devices and transducers. One of the most important points it is necessary to know about all devices is whether they are linear or non-linear, two terms which we must now consider.

Linearity and Ohm's Law

If the voltage in a circuit containing resistance, capacitance or inductance is increased the current will increase. The initial value of the current is determined by the impedance of the combined components and the value of the applied e.m.f. If the voltage is doubled then the current will double; if it is trebled the current will be trebled. This state of affairs is in accordance with Ohm's law which states that the ratio of the voltage to the current in a circuit (a simple one in which the current is driven around by the e.m.f.) is always a constant. It is this constant that we call resistance or impedance. The components in a circuit which obeys Ohm's law are said to be *linear* elements. When certain components, for example valves and transistors, are introduced into a circuit the simple current-voltage relationship expressed in Ohm's Law no longer applies. Because of this such components are called *non-linear*. If the voltage across such a device is increased the resultant current increase does not bear a linear relationship to the voltage change. Amongst the simplest of non-linear devices is a thermionic diode valve. And with this we are back to Edison's experiment mentioned at the beginning of Chapter 1.

Thermionic emission

If a small metal plate is heated with a flame as shown in Fig. 1 (a) its atoms will be excited and the energy of the electrons increased until the energy is sufficient to release them from the parent atom. This phenomenon is rather like the escape of a ball on the end of a piece of string breaking away due to centrifugal force when the ball is whirled around fast enough. The electrons released in this way will drift outside the metal plate, forming a cloud which is quickly diminished as the electrons are collected by the surrounding gas (air). These electrons, collected by the gas, are eventually returned to the metal plate.

If the plate is placed in a vacuum inside a glass envelope, as shown in Fig. 1 (b), the electrons are no longer collected since there is no gas and so a permanent cloud of electrons is formed. The density of this cloud is determined by two factors: the ability of the metal to give off electrons, that is, its effectiveness as an electron emitter; and the temperature of the plate. The higher the temperature, the denser the electron cloud. In practice the metal plate, which is called *cathode*, is made of a nickel alloy coated with a metallic-oxide such as barium oxide or strontium oxide. These oxides are rich in free electrons and when coated on to a material which has the right mechanical and thermionic properties, such as nickel, good emissive characteristics are produced.

In most types of valve the cathode is heated by a filament of tungsten wire wound near to the cathode but electrically isolated from it as shown in Fig. 1 (c), this being the most efficient way of heating a cathode in a vacuum. A current called the filament current or more commonly the heater current is passed through the wire which, because of its resistance, becomes hot. The heat from the filament is radiated towards the cathode which is thus raised to its 'working temperature'. Cathode temperature could be varied by altering the amount of current flowing through the filament but, in practice, this current is fixed.

The diode

If a second nickel alloy plate—or *electrode*—is placed in the vacuum close to the cathode it will collect some of the electrons

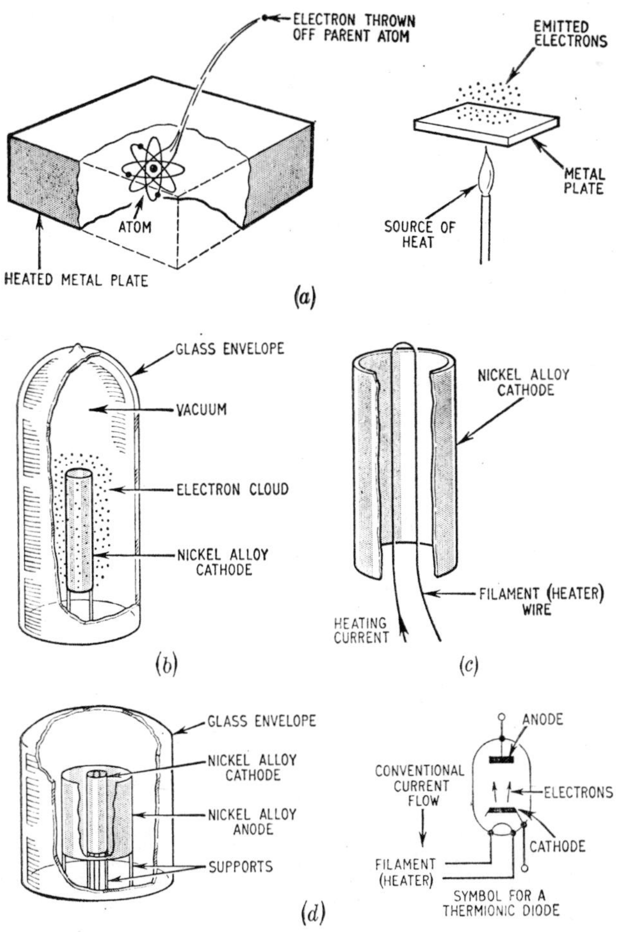

Fig. 1. Thermionic emission.

emitted by the cathode. This process will however continue for only a short while. The cloud outside the cathode represents an area of negative electricity and since the second plate when first placed in the valve envelope is neutral it will only attract a few electrons from the cloud. Then after acquiring a few electrons it will repel any more trying to join it from the cloud, because it will then have attained the same potential as the cathode. This second plate which collects the electrodes is called the *anode*. The flow of electrons from the cathode to the anode is effectively an electron current. Conventional current (see section on batteries in Chapter 2) is assumed to be flowing from anode to cathode, as shown in Fig. 1 (d).

Maintaining a diode current

The arrangement of a cathode and an anode in an evacuated glass envelope forms a thermionic diode valve. There are usually four connections to such a valve, made through the base of the envelope. These are the anode, cathode and the two heater connections. Now if the anode is connected externally to the positive terminal of a battery whose negative terminal is connected to the cathode, as shown in Fig. 2, an electric current will pass through the valve and

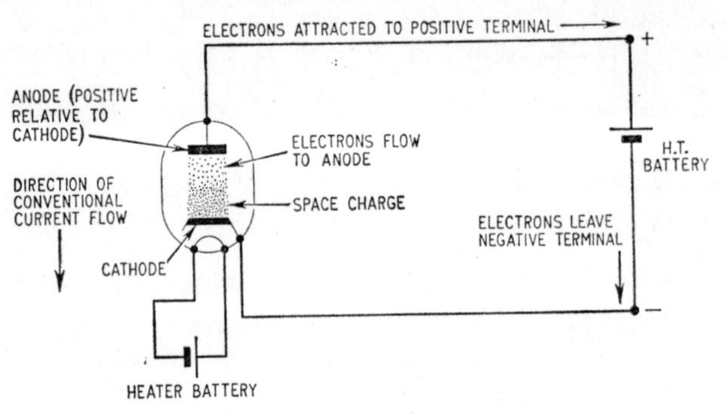

Fig. 2. Thermionic diode circuit.

it is said to *conduct*. The anode, being positive, attracts electrons away from the electron cloud. These electrons, instead of remaining on the anode to make it negative, are attracted towards the positive terminal of the battery. Thus the anode remains positive all the time so that electrons are continuously drawn to it. When they reach the battery terminal the electrons are exchanged inside the cell and at the negative terminal equal numbers come out and travel to the cathode of the valve. Here they are released again by thermionic action to 'refill' the electron cloud. In this way a continuous flow of current is maintained.

The diode conducts an electron current in one direction only, from cathode to anode (that is a conventional current flowing from anode to cathode). The valve will not conduct in the opposite direction because if the anode is made negative it will repel the free electrons. The diode thus behaves as though it were a switch putting in a high resistance to the flow of current in one direction and a low resistance in the opposite direction.

If the voltage—in the correct polarity—between the cathode and the anode is increased then the current through the valve will also increase. The increase of current with increase in voltage is not, however, linear. If, for example, the voltage is doubled, the current may not be increased by more than say one and three-quarters of its original value; if the voltage is trebled the current may not increase more than two and a half times and so on. This non-linearity is a useful characteristic which can be made use of in a number of ways. The thermionic diode is therefore a *unidirectional* and non-linear component.

Using a diode

Perhaps the most common use of the unidirectional property of the diode is in power supply units which take an a.c. mains supply and convert it into a direct current suitable for receivers, amplifiers and so on. If the a.c. is applied to the diode's anode, the output at its cathode will be a series of positive pulses: the diode, being unidirectional, conducts on the positive half-cycles of the a.c. supply only. By connecting a suitable network of capacitors and inductors or resistors in the diode's output circuit these pulses can be smoothed out to provide a steady direct current. This process of converting a.c. to d.c. is called *rectification.*

Thermionic diodes vary between those constructed to handle large currents of the order of several amps and miniature types handling only a few microamperes. Challenging them today in electronics are the semiconductor or solid state diodes.

Semiconductor diodes

Two of the elements used in the construction of semiconducting devices are germanium and silicon. In their pure state these are not electrically suitable but if a small, but controlled, amount of certain impurities is introduced into them then their electric characteristics are improved. Two of these impurities (or *doping agents*) are arsenic and indium. If arsenic is used the semiconducting material is called an n-type material and this under certain conditions will contribute electrons. If the impurity is indium the material is called p-type material and this under certain conditions contributes 'holes' to the action.

The reader has so far been concerned with the movement of electrons and positive ions (in batteries) and the concept of a 'hole' may at first seem difficult to understand. A hole is a space left by the loss of an electron and, whilst it exists, it will always be glad to accept an electron.

A hole can therefore be considered as being the opposite in sign to an electron but having the same mass, that is virtually none. It thus differs from a positive ion, which is relatively very heavy, and can be easily influenced by external magnetic fields.

The pn junction semiconductor

If two pieces of semiconductor material are joined as shown in Fig. 3 a pn junction is formed and this has a significant electrical characteristic. As shown in (*a*) some of the electrons and some of the holes have enough energy to drift over the junction to form a

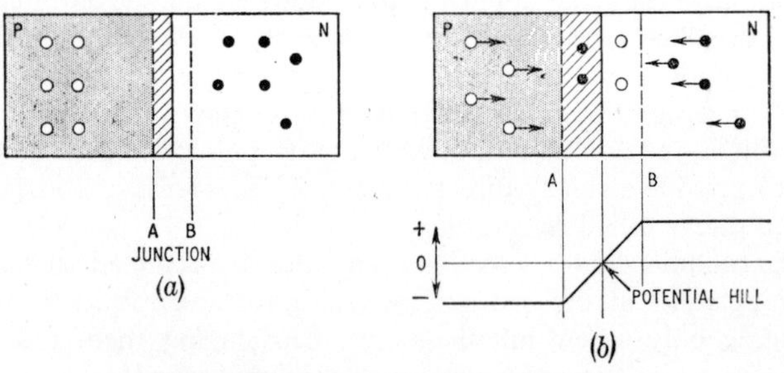

Fig. 3. The pn semiconductor junction.

depletion layer on either side of the junction. This layer has a 'potential hill' which prevents the drifting of further holes and electrons as shown in (*b*). It is rather as though a small battery has been 'built in' to the junction.

Applying bias to a semiconductor diode

Reverse bias.—An e.m.f. applied as shown in Fig. 4 (a) is said to be reverse biasing the junction. The potential hill is *increased* and the drift of holes and electrons is reduced or prevented.

Forward bias.—If an external battery is connected to apply an e.m.f. to the junction as shown in Fig. 4 (b) the potential hill or gradient is *reduced* so that more electrons and holes drift across the

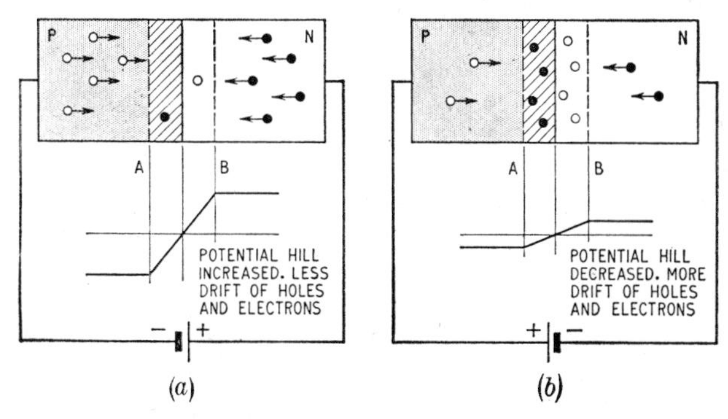

Fig. 4. Biasing a pn junction. (a) Reverse biasing, no current flows. (b) forward biasing, current flows.

junction and a current flows. This is called forward-biasing the pn junction.

Thus the semiconductor diode behaves in the same way as the thermionic diode in allowing current to flow in one direction only. Like the thermionic diode it is unidirectional and non-linear, and it can be used for the same purposes.

This principle forms the basis of many semiconductor arrangements, both germanium and silicon types. Semiconductor diodes are often preferred to thermionic ones because they require no heater and thus no unwanted heat is produced. Equipment, for example a computer, can be made smaller in consequence since no special measures are needed for cooling. However, since semiconductor devices are liable to damage from small increases in temperature the larger semiconductor diodes need a heat sink to prevent them being damaged by the heat produced by the forward current flowing through the resistance of the component. Such an arrangement is shown in Fig. 5.

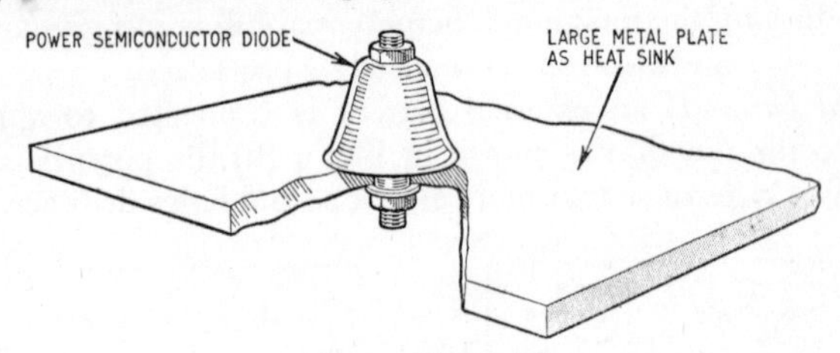

Fig. 5. Heat sink.

Semiconductor diodes are used in many applications, from computers to stabilized power supplies, and their use is steadily increasing. They have high mechanical and thermal reliability and their reverse current (a condition equivalent to making the anode of a thermionic diode negative) is less than with most thermionic diodes. Some typical semiconductor diodes are shown in Fig. 6 together with the symbol used to refer to them in circuit drawings.

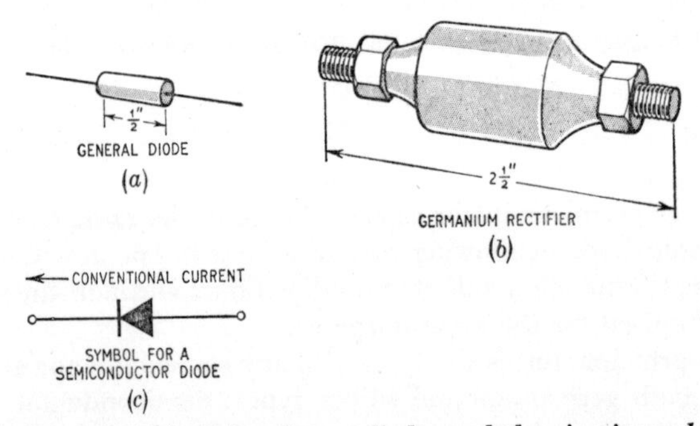

Fig. 6. Two typical semiconductor diodes and the circuit symbol for a semiconductor diode (also used to denote a metal rectifier).

Zener diodes

If a semiconductor diode is reverse biased very little current flows—this being one of their principal advantages. However, if the reverse voltage is further increased an effect known as the zener breakdown occurs. This is shown in Fig. 7.

Under these conditions the reverse current suddenly increases from almost zero to a relatively high value limited only by the resistance present in the external circuit. Changes in voltage across

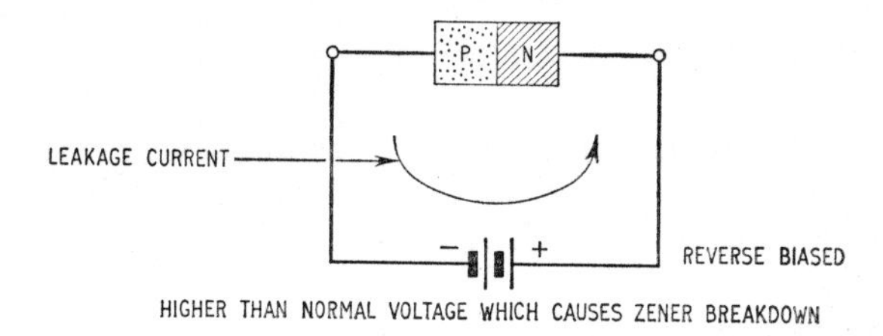

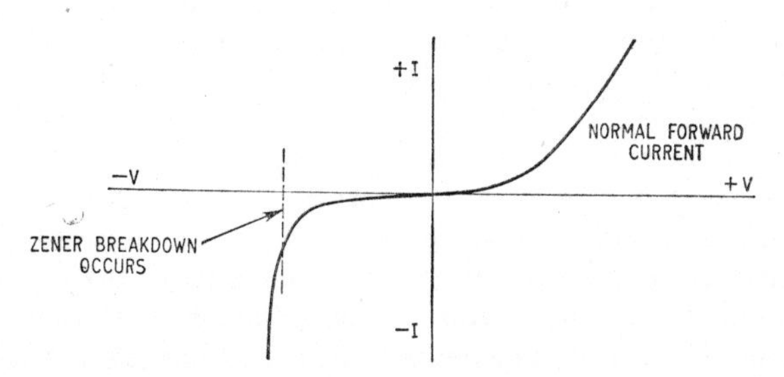

Fig. 7. Action of a zener diode.

the diode have no effect on the value of this current and hence it is very useful as a voltage stabilizing device.

By judicious mixtures of semiconducting materials and doping agents it is possible to fix the value of reverse voltage at which the zener breakdown occurs. Diodes made from such a mixture are called zener diodes.

A zener diode connected as shown in Fig. 8 will stabilize the voltage across AB provided the current drawn by the load is small.

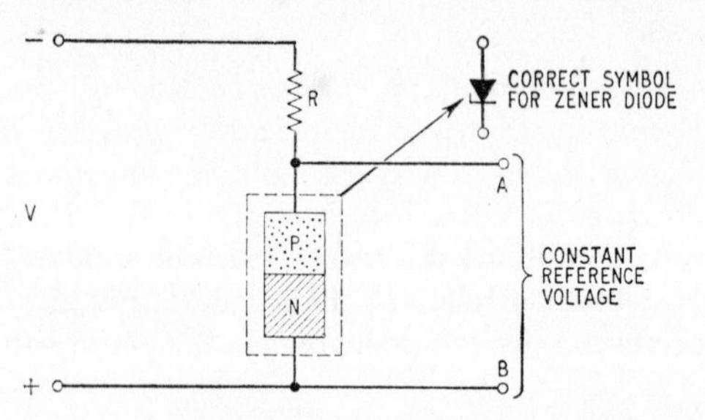

Fig. 8. The zener diode used as a voltage reference source.

In this way it can be used as a reference source of voltage and may, in the near future, replace the standard battery cell as a voltage reference source in electronic equipment.

Tunnel diodes

It is only a few years since Esaki announced his discovery of the tunnel diode. Since the announcement a great deal of research and development work has taken place.

The Esaki or tunnel diode consists of a pn junction of heavily doped p-type material and a similar n-type material.

This heavy doping alters the normal diode characteristics and produces a negative resistance portion on the forward biased current/voltage characteristic. As the voltage across the junction is increased the current increases almost linearly until a point of saturation is reached. The characteristic then flattens and falls off so that further increases in applied voltage result in a decrease in current. This effect represents a negative resistance (since current decreases when the applied voltage increases) and makes the tunnel diode a very interesting semiconductor device. See Fig. 9.

The tunnel diode can be used in many basic circuits which could be applied to such functions as decade scaling, binary counting and other basic computing tasks. Used in conjunction with other micro-miniature components, they may be made as parts of integrated solid circuits (see Chapter 11) to reduce the size of and to aid in the production of very fast high capacity computers of small size.

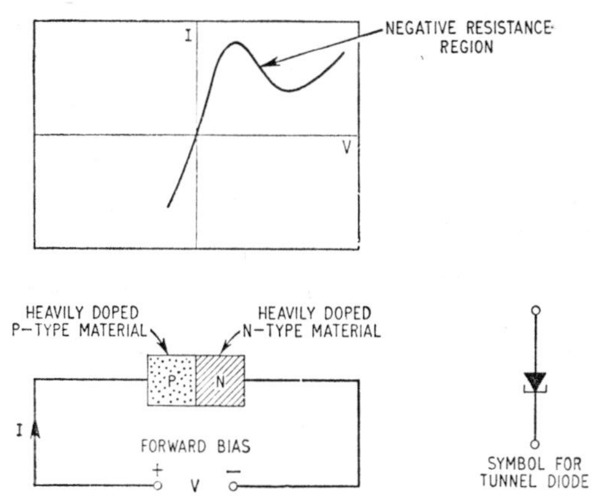

Fig. 9. The principle of the tunnel diode.

Metal rectifiers

Mention should be made here of a further component which acts as a diode, the metal rectifier. This is, like the semiconductor diode, a 'junction' device, but in this case the junction is made between various metals. Copper and selenium are the materials used as the basis of most metal rectifiers. The same circuit symbol as the semiconductor diode one is used in circuit drawings to refer to metal rectifiers.

The need for amplification: the triode

Diodes, both thermionic and semiconductor, perform many essential jobs in electronic circuits. These include detecting, switching, rectifying and so on. If amplification of a signal, a basic function in electronics, is required a more complex component than the diode is needed.

The need for amplification soon becomes apparent to the beginner in electronics. The signals with which electronics is concerned are invariably alternating or pulsed in nature and may be generated at the beginning of a system which could be a television network or G.P.O. link or they might be produced by some action in an automation control process or even generated by the human body in a medical electronics situation. These signals may become attenuated while being passed through various electronic circuits and of course they may be very small in the first place. Before they can be used to indicate that something has happened, or to entertain someone, they must be increased in size and power.

An initial signal voltage must often be increased to perhaps a million times its initial value and the power then increased so that an output voltage is available capable of producing a reasonable current flow to operate a device such as a relay or loudspeaker.

The triode valve

The triode is similar in many respects to the diode except that it has a third electrode inserted between the anode and cathode. This third electrode, called the control grid, is shown in Fig. 10. It is made of a fine nickel alloy mesh and theoretically does not interrupt

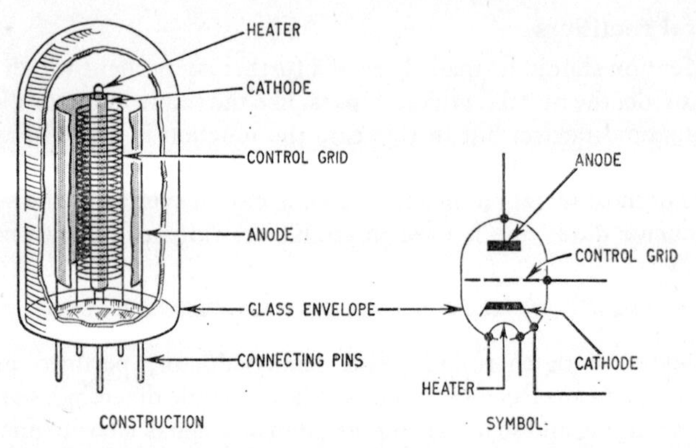

Fig. 10. Miniature thermionic triode—construction and circuit symbol.

the flow of electrons from the cathode to the anode as they pass through its mesh. In practice it does tend to collect a few of the electrons and so it very slowly acquires a negative charge which begins to repel the electrons from the cathode and would eventually reduce the current through the triode to zero. To avoid this it is usual to connect the control grid through a resistor back to the cathode to allow the collected electrons to drain away. This resistor is called the 'grid leak' resistor.

Applying a signal

Suppose that a voltage is applied between the control grid and the cathode. What will its effect be?

If this voltage is negative with respect to the cathode, as shown in Fig. 11 (a), then it will repel electrons from the cathode so that only the more energetic will pass through the mesh to the anode. The current through the valve will have been greatly reduced.

If the control grid is made positive with respect to the cathode, as shown in Fig. 11 (b), the grid will act like the anode and attract electrons but, because of the grid structure, most of them will pass

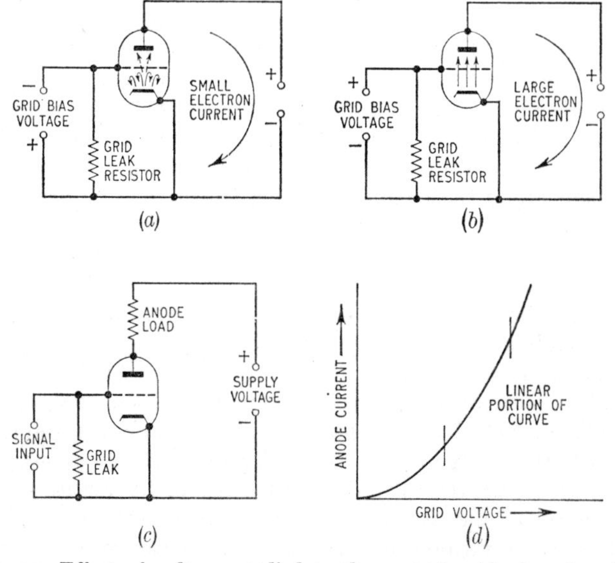

Fig. 11. Effect of voltage applied to the control grid of a thermionic triode. (a) Negative bias: small anode current because most electrons are repelled. (b) Positive bias attracts electrons so that the anode current is large. (c) Basic triode amplifier arrangement. (d) Typical grid voltage/anode current characteristic curve.

through the mesh and go to the anode to be collected. As a result, the current through the valve will have been greatly increased.

Thus it can be seen that the anode current (the current through the valve) can be controlled by a voltage on the control grid. It is possible to control relatively large currents through a valve with small voltages, perhaps only millivolts, on the grid.

The grid is able to exercise this control because it is physically nearer to the cathode than is the anode and so a small voltage has as much effect at this point as a considerably larger voltage on the anode. The fact that the grid is a wire mesh enables it to create attracting and repelling electric fields without actually collecting many electrons (so few as to be theoretically regarded as zero).

The voltage applied to the grid does not need to supply any current in theory because none flows in the grid circuit and this is especially advantageous because there is not usually any power to spare in the input signal (when an applied voltage makes current flow power is dissipated).

Suppose now, as shown in Fig. 11 (c), a resistor, called the *anode load resistor*, is inserted in the anode lead to the valve. As the varying signal input voltage varies the current flowing through the valve and round the circuit, a varying voltage will be developed across this anode load resistor. This variation in voltage will be considerably greater than the initial signal voltage variation, and thus the valve acts as an amplifier.

We mentioned earlier that the valve is a non-linear device. That is, if a graph is drawn of the grid voltage plotted against the anode current, as shown in Fig. 11 (d), it will be seen that the result is curved. Because of this, the voltage appearing in the anode circuit across the anode load resistor may not be an exact amplified replica of the voltage at the grid, and because of this *distortion* is said to occur, i.e. the valve distorts the signal. By careful design and control of the circuit conditions, however, the grid voltage-anode current relationship can be made linear over a certain portion of the curve, so that, provided the signals are kept within this range, the distortion can be kept to a negligible amount. A curve such as that shown in Fig. 11 (d) is called a valve's *characteristic*. In the section on amplifiers in Chapter 4 the means—biasing

—by which a valve is made to operate on the straight part of its characteristic is described.

Pentode valves

The triode is a much used valve in electronics. In some applications, however, its performance is limited and a pentode valve is employed instead. This valve has two further grids as shown in Fig. 12. The first one, called the screen grid, is generally connected to a positive voltage and, via a large capacitor, to the cathode. This grid is inserted between the control grid and anode in order to 'screen-off' the unwanted capacitance which exists between the

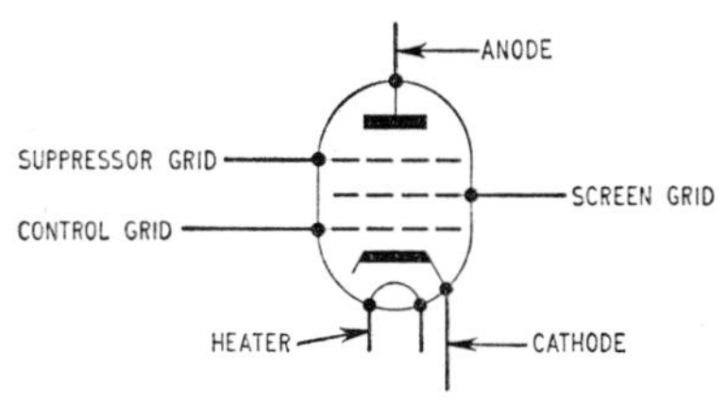

Fig. 12. Electrodes of a pentode valve.

anode and the control grid and is troublesome in some applications. The screen grid on its own also, however, produces a characteristic which has certain disadvantages and to overcome these a third grid, called the suppressor grid, is inserted. This is generally connected to the cathode.

The pentode has many interesting characteristics—one of which is the ability to produce a constant current whatever the voltages on its grids (within limits).

The transistor

The transistor is the semiconductor equivalent of the triode valve. It is made by using three 'slices' of semiconductor material. A slice of n-type material may be sandwiched between two slices of p-type material to give a pnp transistor or a slice of p-type may be sandwiched between slices of n-type material to give an npn transistor. In the transistor there are two junctions and control is achieved by a varying signal applied to one section of the transistor.

A typical transistor of the npn junction type is shown in Fig. 13 together with the symbol used for a transistor in circuit diagrams. The transistor may be made of either silicon or germanium.

As shown in the diagram the three sections of the transistor are named the emitter, base and collector, the base being in this case the common terminal so that the other sections can be biased in some way.

In the arrangement shown in Fig. 13 the emitter-base section is forward biased and the collector-base section reverse biased. Under these conditions any change in the emitter current produces a

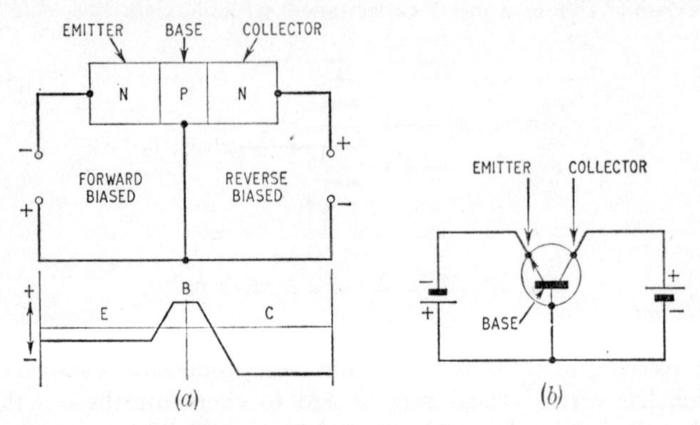

Fig. 13. The npn transistor.

change of a similar amount to the collector-base circuit current because it allows more electrons or holes to drift across the junction. It is exercising control like the control grid in a triode valve.

Since the impedance of both the emitter-base and collector-base circuits are different some power gain is possible.

This can be better illustrated by looking at Fig. 14 (c) which shows a transistor connected as a 'black-box' with two input terminals and two output terminals. In the arrangement shown the input impedance is much lower than the output impedance. If the current in the input circuit is changed, say by a few milliamperes, then a small amount of power, the amount being determined by the impedance in the circuit, is used—dissipated. This variation in

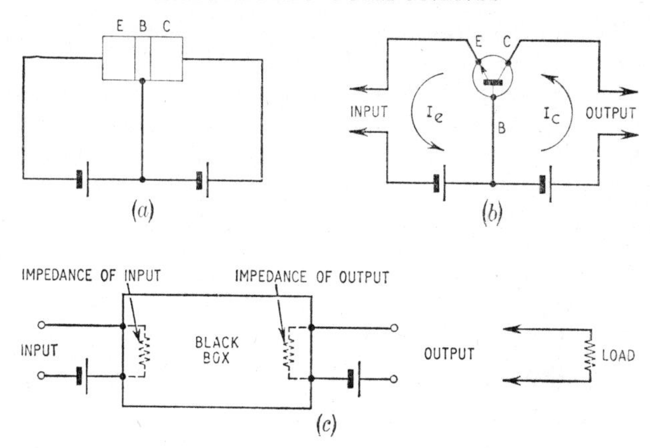

Fig. 14. The transistor as a 'black box'.

current in the input stage causes the current in the output stage to change by a similar amount. However, as the impedance in the output circuit is larger so the power which is dissipated is greater. Between input and output circuits, therefore, we get some power amplification. If the input and output circuits are matched then the maximum power will be fed into the input circuit from the source, amplified and the maximum amount at the new amplified level fed out to the load—which may be an indicator or a loudspeaker, etc.

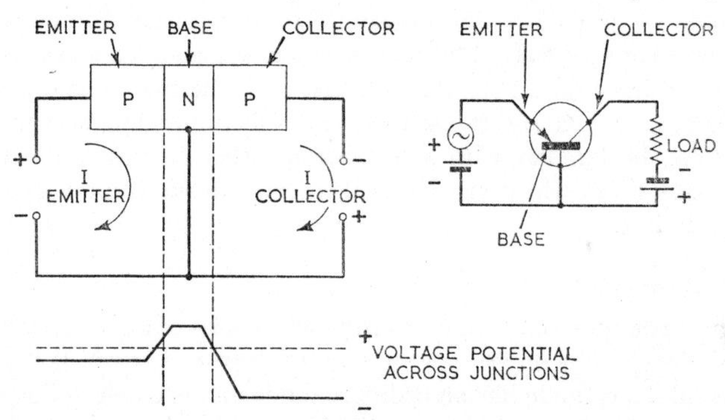

Fig. 15. The pnp transistor.

In this way the transistor can be used as a power amplifier and has a similar function to a triode valve.

The pnp transistor is shown in Fig. 15, and it will be seen from this that the polarity of the supply voltages are the opposite to those of an npn transistor, i.e. the pnp transistor requires a negative voltage at its collector while an npn transistor requires a positive voltage at its collector.

The transistor may be connected in circuit in any of the three ways shown in Fig. 16. There are alternative names for each

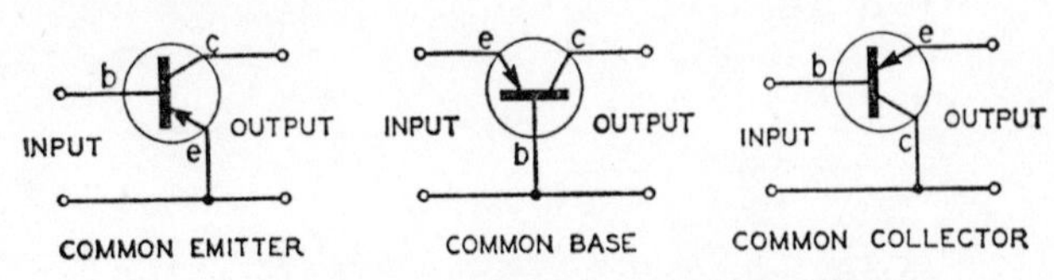

Fig. 16. Three basic ways of connecting a transistor.

arrangement. The common emitter is also known as the grounded emitter, the common base as the grounded base, and the common collector as the grounded collector or emitter follower.

The subject of transistor amplifiers is taken further in Chapter 4.

SOME OTHER IMPORTANT COMPONENTS

So far we have discussed some of the most important components in electronics—linear devices (resistance, capacitance and inductance) and non-linear devices (diodes, both thermionic and semiconductor, triodes and transistors). The electronics engineer must, however, be familiar with a number of other electronic devices, especially those which can be used to convert one form of energy into another.

The neon tube

The neon tube is shown in Fig. 17 and consists of a glass envelope filled with pure neon gas or a mixture of neon and argon. It has an anode and a cathode like an ordinary diode and is connected in the same way. If a low voltage is applied across the electrodes nothing will happen, no current will flow. As the voltage is increased, however, a certain point is reached, depending upon the characteristics of the tube, when the tube will 'strike'. Current suddenly begins to flow and the fast-moving electrons knock more electrons off the

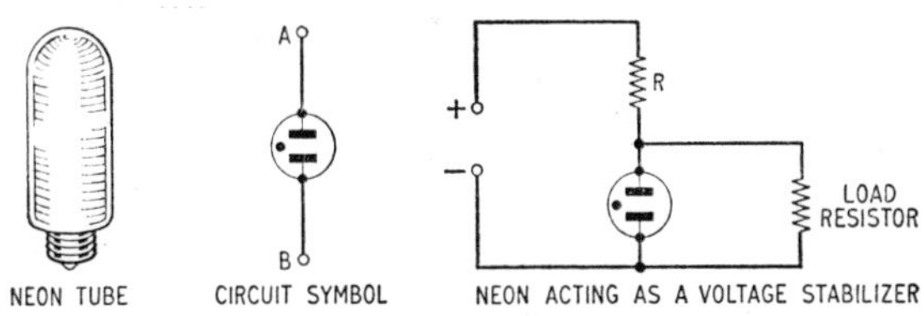

Fig. 17. The neon tube—appearance, circuit symbol and use as a voltage stabilizer.

gas molecules in the tube ionizing them. There is a rapid increase in current and the resistance between the anode and cathode suddenly changes from a very high value to almost zero. The tube can be seen to be 'struck' because the ionized gas glows inside the glass envelope.

The neon has many uses but the most common is as a voltage stabilizer—rather in the same way as the zener diode.

The thyratron

A triode valve filled with a gas like argon, neon or hydrogen is called a gas-filled triode or thyratron and has a useful role in electronics as a switching device.

If the grid is biased sufficiently negatively relative to the cathode then no current flows between the anode and cathode and the gas in the tube remains de-ionized. If, however, the grid voltage is suddenly made positive (e.g. by the application of a positive-going spike waveform) the thyratron begins to conduct. The sudden movement of electrons in it produces many collisions with the molecules of the gas resulting in even more free electrons being produced and accelerated between cathode and anode. The result is a sudden and big increase in anode current. The significant factor about the thyratron is that when this occurs the grid no longer has any control and the anode-cathode current will flow for as long as the anode voltage remains above a certain value. (In practice it is usually arranged that it falls fairly soon after the event starts.)

A thyratron's usefulness is that it will produce a fat, powerful square pulse from a less powerful short duration spike.

The silicon controlled rectifier

A modern semiconductor device that does the same job of work as the thyratron is the silicon controlled rectifier. It consists of two p-type sections of silicon and two pieces of n-type silicon arranged alternatively as shown in Fig. 18. With a d.c. voltage connected in the manner shown at (a) the junctions 1 and 3 have a forward bias condition and junction 2 is reverse biased. The behaviour of the whole component is governed by this reverse bias junction which

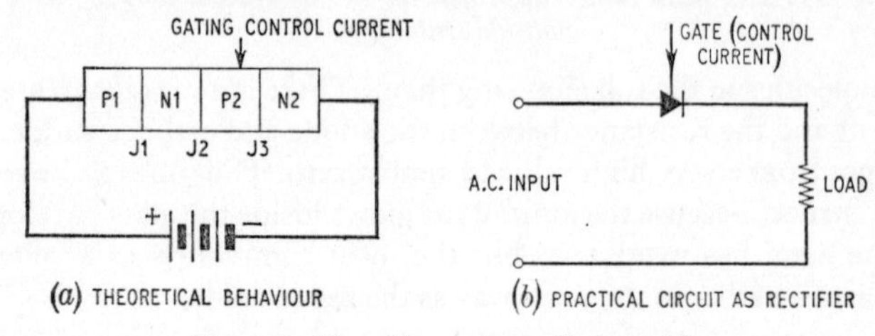

Fig. 18. The silicon controlled rectifier.

has a 'gate' connection at P2. The application of a pulse to the gate connection modifies the behaviour of the junction 2, producing a breakover effect so that the junction suddenly behaves as though it is forward biased. The whole component now acts as a forward biased pn junction and a large current flows. Like the thyratron, control is now lost and the current can only be reduced by means external to the device. The silicon controlled rectifier can be considered as a current-operated switch whereas the thyratron is a voltage-operated switch.

Many of the jobs formerly done by thyratrons are today done by these 'solid state' rectifiers.

Thermistors

The thermistor enables heat energy to be changed into electrical energy and is therefore very useful for measuring temperature. It is usually made by taking a mixture of nickel, cobalt, manganese and other oxides, forming them into a small bead and inserting two connecting leads. The unit is then fired to give a ceramic-like appearance.

As the temperature surrounding a thermistor increases so its resistance decreases (that is it has a negative temperature coefficient). This change can be detected electrically and displayed on a meter. Fig. 19 (a) shows a typical thermistor and the circuit symbol and

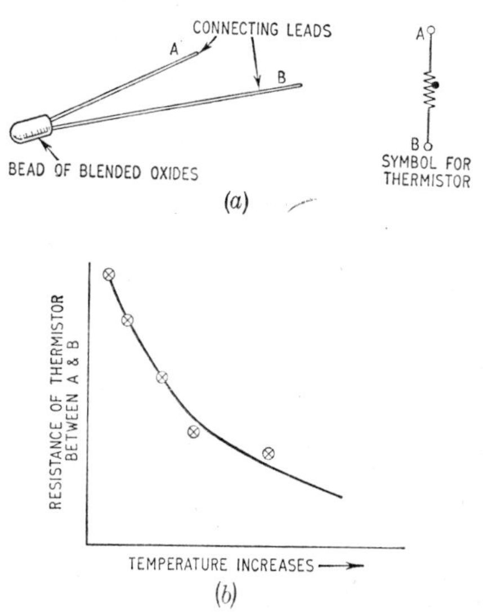

Fig. 19. Typical thermistor and its temperature/resistance characteristic.

(b) shows a typical relationship between its temperature and resistance. Thermistors are commonly used in circuits as compensating devices, correcting the behaviour of a circuit when the temperature alters.

Transducers

Transducers are devices which convert mechanical energy into electrical energy or vice versa. They take many forms, for measuring velocity, acceleration, pressure, torque, distance, and so on. An example is shown in Fig. 20 (a). It consists of a resistance made of a coil of resistance wire with a variable tapping attached to it: any movement of this tapping by mechanical means alters, as can be seen, the amount of resistance in circuit. The change is detected by measuring the current flowing through the varying resistance

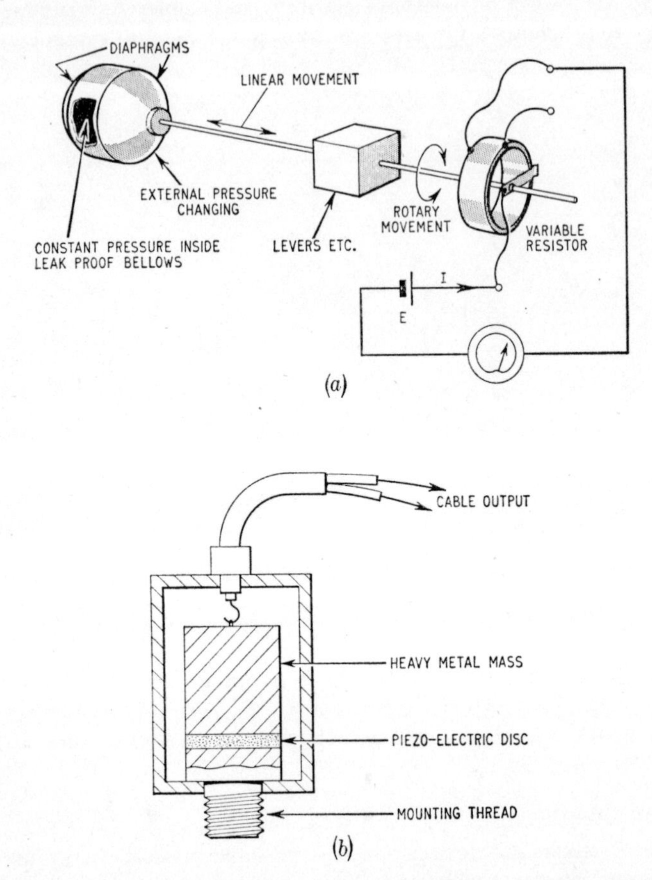

Fig. 20. (a) Simple transducer for measuring pressure. (b) A piezo-electric transducer designed for use as an accelerometer in missile research.

in the circuit produced by a constant e.m.f. (the battery). This elementary form of transducer and variations of it are used in many applications where mechanical movement must be accurately measured.

Pressure can be measured by using a small pair of bellows whose diaphragms move with pressure change and move a variable resistance or alter the tuning of a circuit having capacitance and inductance so that the pressure changes become changes in electrical currents.

Piezo-electric material is now used for many transducers. This type of material, often lead-zirconate or barium titanate, is capable of providing an e.m.f. when it experiences any mechanical force. In the same way if electrical energy is applied to the material it will exert a mechanical force.

There are numerous applications of transducers made from piezo-electric materials—they can indicate pressure, acceleration or produce music when they are used in gramophone pickups. In ultrasonics, high frequency electrical currents are used to energize the transducer and produce correspondingly high frequency pressure vibrations which are used to do many jobs from detecting submarines to cleaning glassware. The piezo-electric transducer can be used either as a transmitter or receiver of mechanical energy.

In Fig. 20 (b) a piezo-electric transducer of the type used to measure acceleration in missiles is shown. Sudden change in pressure on the piezo-electric material produces an e.m.f. proportional to the acceleration.

The use of ceramics for piezo-electric devices enables transducers to be made in almost any shape with any axis of activity according to requirements. (At London University a ceramic crystal of special type has been adapted for making delicate measurements in studying the movements of amoeba). The axis of activity of the crystal is decided and the ceramic shape fitted into a jig and immersed in special insulating oil. Across the chosen plane a polarizing field of several kilovolts per millimetre is applied for several seconds, at high temperature. The combination of high temperature and electric field reorientates the molecules in the ceramic to create the piezo-electric properties of the material.

Photo-electric devices

These devices convert light energy into electrical energy and are used extensively in automation engineering. In Fig. 21 a photo-electric cell using selenium as its active layer is shown. When light falls on the cell it penetrates the transparent layer, which is made of metal oxide (not glass as may be supposed). Since it must be

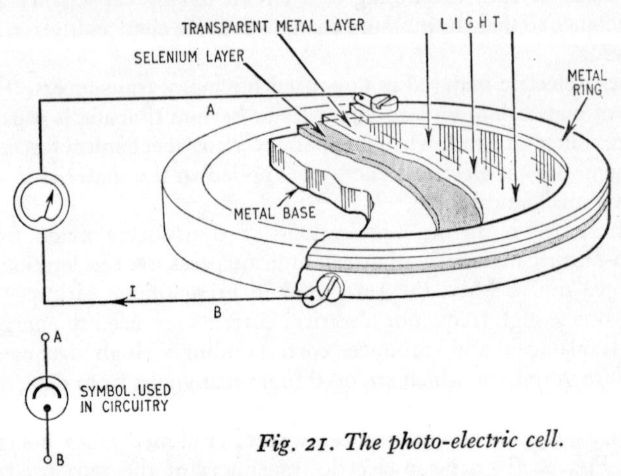

Fig. 21. The photo-electric cell.

electrically conducting, evaporated stannous oxide can be used for this purpose. The light passing through acts upon the selenium layer and releases electrons in proportion to the amount of light falling on the sensitive part of the cell. The freed electrons, called photoelectrons, form an electric current which can be used to indicate directly the amount of light falling on the cell.

The dekatron

A very useful component which is frequently encountered in electronics is the dekatron counting tube. In Fig. 22 the tube is shown diagrammatically. All the guide electrodes immediately after the cathodes are connected together to form the 'A' group of guide pins, and likewise all the other guide pins are connected to form the 'B' group. Both groups are connected to the signal input, a series of pulses (which must be negative going) that are to be counted, but the 'B' group is connected to the input through an electrical delay unit. This means that the pulses arrive at the 'B' pins a fraction of a second after they reach the 'A' pins. All the cathodes, which represent numbers 0 to 9 on the face of the tube, are connected together and taken to the negative side of a d.c. supply source, the

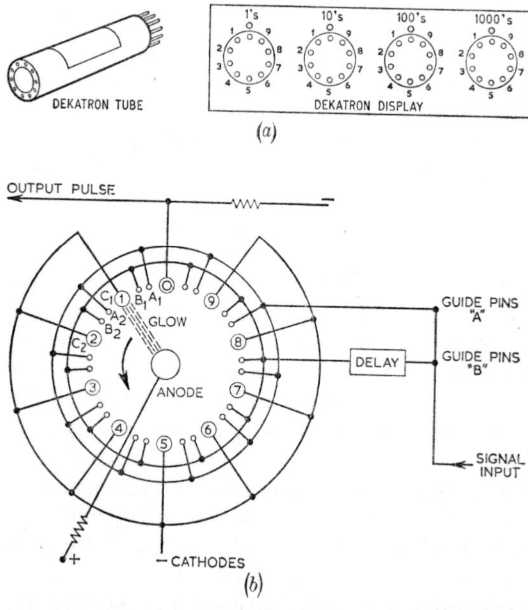

Fig. 22. The dekatron. (b) Electrode arrangement of the dekatron.

single anode being connected to the positive side. Suppose the tube has struck (rather like a neon) between the anode and cathode 1 as shown. This means that one pulse has been counted and a second is due to arrive. When it does arrive the guide marked A2 goes negative by an amount which is arranged to be slightly greater than the voltage on the adjacent cathode (1). The result is that the 'glow' jumps from the cathode to guide pin A2. A fraction of a second later guide pin B2 goes negative also and the glow jumps from pin A2 to pin B2. The pulse then ceases and A2 and B2 go dead. The glow now jumps to the nearest negative electrode, which is cathode 2. In this way the pulse is counted, having made the glow jump from cathode 1 to cathode 2, the two guide pins assisting in the movement. This process will continue until ten pulses have been counted. At this point the glow arrives at the tenth cathode, marked 0, and it is usual to arrange that this cathode passes on a pulse of its own to another dekatron tube which will as a result count one pulse. For every ten pulses fed to the first tube the second counts one, that is it counts in units of ten. By connecting together a series of dekatrons in this way it is possible to count up to millions.

Relays

The relay is another important component. It is, in a sense, a transducer since an electrical current applied to it produces a mechanical movement. Invariably however this movement is immediately converted back into an electrical change because the relay is primarily used to switch other circuits on or off.

In its simplest form the relay consists of a coil of wire wound on a piece of soft iron. When an electric current is passed through the coil the magnetic field created makes the soft iron into a temporary magnet. This 'temporary magnet' is so arranged that when it is magnetized it attracts blades of electrical contacting material in such a way as to make—or break—one or more other circuits, rather like an automatic switch. As the current flowing in these other circuits may be very considerably larger than the current required to operate the relay, the usefulness of the device is its ability to control large currents by small ones.

By using fast acting relays many useful control circuits can be devised. Perhaps one of the most interesting of modern fast-acting relays is the dry-reed relay which only made its appearance on the commercial market in its present useful form comparatively recently. This type of relay is a miniature, fast-acting one which has almost no *contact bounce* (the effect, obviously undesirable, of the contacts bouncing on and off due to mechanical shock when the relay acts). In the dry-reed relay two reeds of nickel-iron are placed in a small glass envelope which may be less than one inch in length. The envelope is evacuated and the leads to the reeds brought outside for external connection. The tips of each reed are coated with gold or some other suitable material. Around the glass envelope a coil is wound and through this is passed the operating current. The current through the coil creates a magnetic field axially along the length of the glass envelope, and this field makes the reeds snap together giving a fast contact without bounce. The time taken for this to happen may be as low as a millisecond. The component is illustrated in Fig. 23. As the contacts are in an evacuated envelope they are free from contamination and many of the other adverse

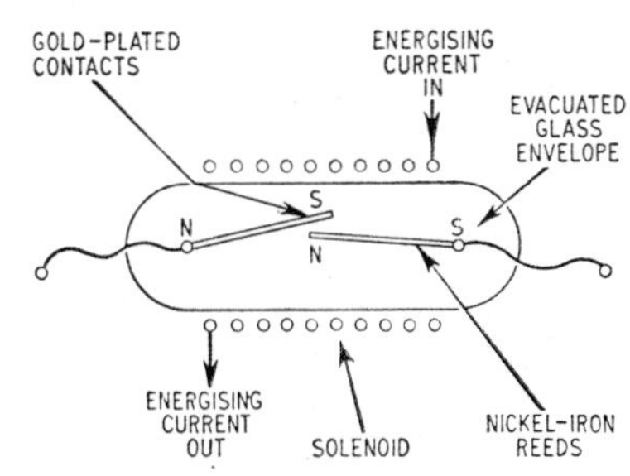

Fig. 23. Dry-reed relay.

conditions affecting conventional relays. And because of the mechanical simplicity of the system the dry-reed relay is an inherently reliable component.

Semiconductor devices, e.g. transistors, can also be used as fast switches (they are often referred to in this connection as *static switches*) and at present there is much competition between relays and semiconductor switches. (A conducting semiconductor device is the equivalent of a closed switch whilst a non-conducting semiconductor device is the equivalent of an open switch.) The decision as to which to use is, however, really a matter of carefully studying the particular application to decide which is best suited. Even the fastest ordinary relay is much slower in operation than a semiconductor switch circuit but, on the other hand, the relay has the advantage that it can do a more complex job.

4
BASIC ELECTRONIC CIRCUITS

The components which we have been discussing in the previous chapter can be arranged in various ways in electronic circuits in order to fulfil a number of basic functions. In all electronic operations we are concerned with a signal, which may be an alternating current oscillating at a very high frequency, a series of pulses or spikes, a waveform which varies in amplitude or a direct current which is fluctuating slowly. In some electronic equipment this signal may be fed in from an outside source and require, for example, detection and measurement; in other instances we may have to generate the signal, apply it to some outside process, and then later detect it again in order to see what has happened to it in the meantime. There are a number of basic circuits that play a part in all or a few of the many possible applications of electronics. In nearly every application, for example, we need circuits which will *amplify* a weak signal; in many applications we need various circuits which generate various different types of signals—oscillators to provide high frequency a.c. signals and timebase generators to provide pulses and spikes. Also there are a number of basic circuits using diodes or simple resistance-capacitance networks which we use in order to shape or otherwise modify the signals. In this chapter what can be considered as the important basic circuits are outlined.

SIMPLE AMPLIFIERS

One of the most common operations necessary in electronics is amplification. Signal voltages and currents are often very small initially and, as they pass through equipment, losses which produce *attenuation* of the signal may occur. They must therefore be amplified before they can be useful. The output of a radio communications transmitter, for example, may be very high but several hundred miles away it might only be capable of producing a few millivolts of signal in a radio receiving aerial. This is too small to operate a loudspeaker or recording chart so that amplification is necessary before use can be made of it. In medical electronics the small signals present in the human body are measured and presented as a record to the clinician for his assessment: these signals are usually very small, often only microvolts, and they must be considerably magnified before a full understanding of their significance is possible.

Gain of an amplifier

Engineers talk about the *gain* of an amplifier and this is generally expressed in decibels (abbreviation dB). The decibel is a unit used to compare the ratio of two quantities. Sounds are also usually measured in dB. A road drill, for example, is said to be as noisy as 90 dB. This means that the ratio of the noise of the drill to some other noise, for example, a whisper, when expressed in decibels, equals 90. In amplifiers it is the ratio of the output voltage to the input voltage that is given in dB as the gain (provided that the input

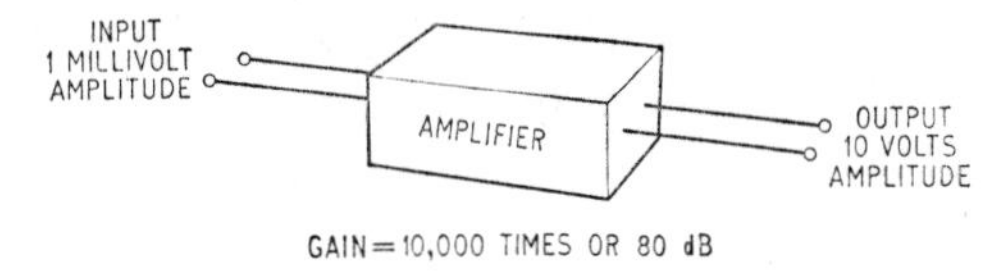

Fig. 1. Amplifier considered as a 'black box'.

and output impedances are the same). Fig. 1 shows a box representing an amplifier. The input voltage is 1 millivolt and the output is 10 volts. This means that the amplifier has amplified 10,000 times or, expressed in decibels, it has a gain of 80 dB.

The useful thing about decibels is that they can be added directly together. For example, if an amplifier with a gain of 50 dB has its output connected to the input of an amplifier with a gain of 100 dB, then the total gain is 150 dB.

In Table 1 a range of decibel values against actual gain is shown.

It can be seen from the table that the dB figures increase more slowly than the other ones. For example, an amplification of 10 times equals 20 dB and yet one of a 1,000 times equals only 60 dB.

It may seem that the dB method of stating gain is a misleading one but in practice it enables the designer to keep a sense of proportion: amplification itself is not enough, it is a good signal-to-noise ratio that counts (in all electronic equipment there is a certain amount of background interference, called *noise*, caused by random electron movements, etc.).

Table 1

V_{out}/V_{in}	Gain in dB	V_{out}/V_{in}	Gain in dB
1	0	80	38
2	6	100	40
4	12	200	46
8	18	400	52
10	20	800	58
20	26	1,000	60
40	32		

A simple single triode valve amplifier, providing voltage amplification, is shown in Fig. 2. If a small signal voltage is applied between the grid and cathode of the triode hardly any current flows in this part of the circuit and the input is said to offer a very high

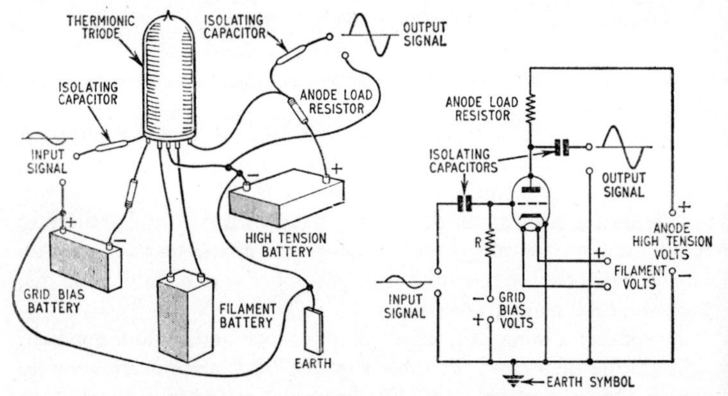

Fig. 2. Simple thermionic triode amplifier. Left, connections between the components; right, circuit diagram.

impedance, an important advantage. The signal, applied through an isolating capacitor, causes the electron stream from the cathode to the anode of the triode to be varied (as outlined in Chapter 3). This varying current flows in the external circuit through the anode load resistor. A potential difference is developed across this resistor, and this is proportional to the value of the anode current and to the value of the resistor. The voltage produced in this way may be a hundred times greater than the voltage at the grid of the valve and follows any variations in the grid voltage. It may seem that by increasing the value of the anode resistor there is no limit to the amount of amplification that can be obtained in this way but, in practice, an upper limit to the value of the anode resistor is set by the valve characteristics and is usually specified by the manufacturer.

A number of amplifying valves can be connected in series, the output of one being connected to the input of the second and so on. This is called *cascade* connection. By doing this a three-stage amplifier, for example, will provide a gain of $100 \times 100 \times 100$, i.e. 1,000,000 times or 120 dB gain. There is, however, an upper limit to the number of stages that can be connected in this way. If too many are connected in cascade minute fractions of the output may unavoidably get back to the input: when this occurs the amplifier is said to be unstable, the amplifier oscillating violently as the energy returns from output to input cyclically.

Valve characteristics: grid bias

Fig. 3 (a) shows the changes that occur in the anode current of a triode valve when various voltages are applied to the grid. The curve derived from this—see Fig. 3 (b)—is called the valve characteristic (valve characteristics were briefly mentioned in Chapter 3). Examining it, we see that the current through the valve ceases when the bias voltage, V_g, reaches about $-7V$. The valve is then said to be *cut off*. If, however, a voltage of $+1V$ is applied to the grid the maximum amount of anode current will flow and further increases in the grid voltage will have no effect on the anode current. This point is called the *saturation point* of the valve. Between the cut-off and saturation points there is a portion of the valve characteristic

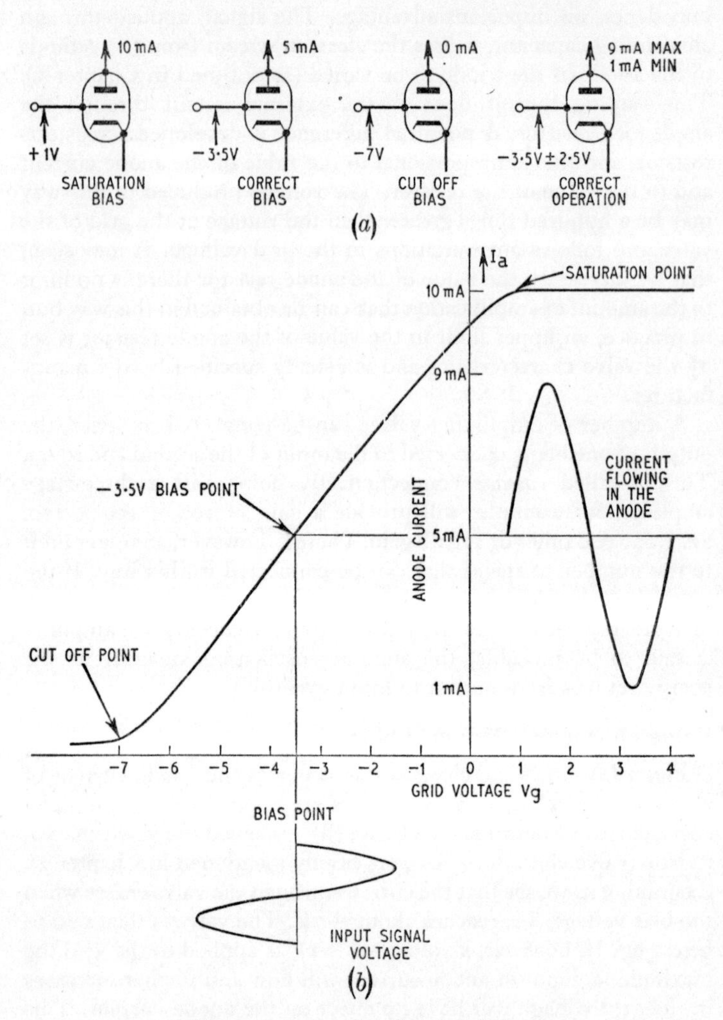

Fig. 3. Triode valve characteristic curve.

that is straight (or linear) and if the grid voltage varies within certain limits on this portion, in this case $\pm 2{\cdot}5V$ on either side of the $-3{\cdot}5V$ point, then the anode current will change in a linear fashion and the output signal voltage will be an undistorted, amplified replica of the input signal voltage, that is it will have exactly the same shape as the grid voltage. The mid-point of the linear portion of the characteristic is called the *bias point* and the control grid of a valve may be supplied with a fixed negative bias voltage so that it operates on the straight part of its characteristic. In the case illustrated in Fig. 3 the bias voltage would be $-3{\cdot}5$ volts.

The grid bias voltage can be applied through a large resistor, in practice the grid leak resistor, because, as no current is drawn from the bias source, there is no loss of bias voltage across it (see Chapter 2). The resistor also prevents the input signal from being shorted out by the bias source.

Cathode bias

In practice it is unusual to use a separate supply, e.g. battery, to provide the biasing voltage for a valve. The desired effect can be achieved in a simpler way. If a resistor is placed in the cathode lead of the valve, then the current flowing through the valve will build up a voltage across this resistor. This voltage will be determined by the current flowing through the valve and the value of the resistor, and will make the cathode positive with respect to earth. If, now, the grid is returned to earth via the grid leak resistor there will be a potential difference between the grid and cathode, that is the grid will be negative with respect to the cathode. Thus placing a resistor in the cathode lead has the same effect as applying a negative bias voltage directly to the grid of the valve. This technique is called *cathode biasing*. To prevent the bias varying in value with the variations in the signal current a large value capacitor is often connected across the cathode bias resistor to smooth out these variations.

The isolating or coupling capacitor

The amplifier we have been discussing is used to magnify alternating or varying signals and therefore it is possible to use capacitors to connect the signal to the grid of the valve and also to take away the output from the anode of the valve. These capacitors isolate the high tension supply voltage at the anode of one valve from the grid of the next stage. The capacitors must be large enough so that their capacitive reactance does not seriously diminish the value of either the signal coming into the amplifier or that being passed on.

The transistor as an amplifier

The transistor can be used as an amplifier and, in the same way, needs to be biased to established the right conditions for amplification. The transistor, however, is a current operated device and is therefore biased by a current, not a voltage. Fig. 4. shows a simple pnp transistor amplifier. The emitter has a positive bias with

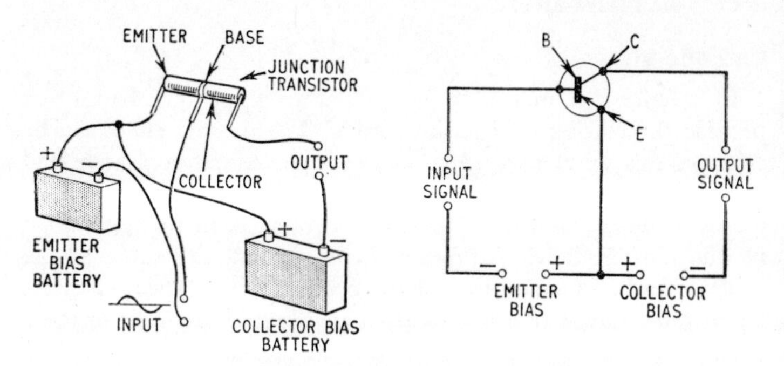

Fig. 4. Simple transistor amplifier.

respect to the base so that holes are able by the modification of the barrier potential to drift across this junction. Due to the thinness of the base the holes drift across to the collector which, since it is negatively biased, collects them. The emitter-base current is very small under these conditions whilst the emitter-collector current is almost equal to the total emitter current. Because the emitter-base impedance is low, a low power is needed in this circuit to control the relatively large collector current. Since the collector-base impedance is high, a large power change occurs which thus represents a power gain. This power gain can, using suitable components, be converted back into a current or voltage gain.

Biasing transistors

The polarity of the d.c. supply and the bias current for a transistor depends on whether the transistor is a pnp or npn type and also on the circuit configuration—whether the transistor is connected in the common base, common emitter or common collector arrangement. In the case of the most frequently encountered circuit, the common emitter with a pnp transistor (Fig. 5 (a)), a negative bias current is required at the base to make the transistor operate at a suitable point on its characteristic for amplification. This, as

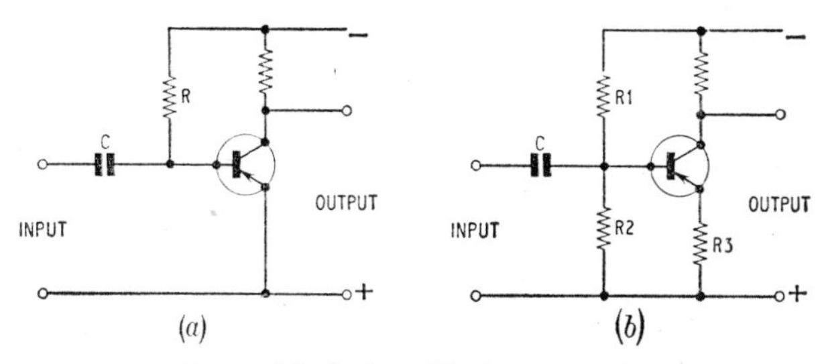

Fig. 5. Methods of biasing a transistor.

shown, can easily be obtained from the negative h.t. line through the resistor R, the capacitor C being an isolating capacitor.

Transistors are particularly sensitive to temperature, which affects a transistor's characteristics. This means that the operating point will move with changes in temperature unless precautions are taken to avoid this. The usual methods of overcoming this are to add resistors between the base and earth (Fig. 5 (b), R2) and between the emitter and earth (R3). This arrangement provides sufficient stabilization for most purposes. Increased stability can be achieved by using thermistors in the circuit.

D.C. amplifiers

The amplifiers so far discussed have been ones for handling a.c. signals or pulses. Altering the voltage on the grid of a valve causes a change in anode current and a change in voltage across the anode load. If this is a slowly varying signal voltage, however, it will not be transmitted through the coupling capacitor. To amplify slowly changing or direct voltages and currents a different technique is needed: direct coupling between stages must be used, these amplifiers being called d.c. (direct coupled) amplifiers. They are used in many applications, especially in medicine, chemistry and automation, and must be capable of magnifying very small and very slow changes in d.c. levels. This is rather tricky because due to the absence of capacitors the high tension from one stage is present on the grid of the next stage. To overcome this a high negative bias voltage is applied to the grid—see Fig. 6—to overcome the high

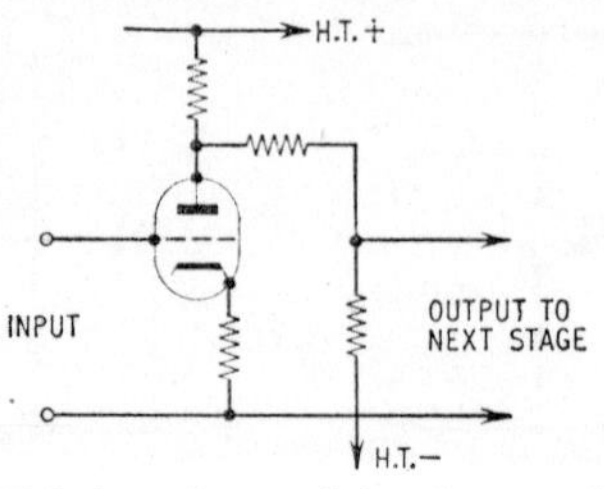

Fig. 6. Triode valve used for d.c. amplification.

positive voltage at the anode of the previous stage. But any small drift in the bias conditions and high tension will cause drifting which is felt throughout the system at a spurious signal. In some d.c. amplifiers the first stage, which is the most sensitive to drift, is designed with an input using two valves arranged so that the drift effects in each valve cancel out. This is called a *balanced* input since both grids are at the same voltage with respect to earth.

Transistors, which do not need h.t. and heater supplies, are a considerable improvement for d.c. amplifiers. In particular using npn and pnp transistors alternatively reduces drift effects.

Output stages

So far we have been considering voltage amplifiers, which are mainly found in the preliminary stages of multi-stage amplifiers. The output of the multi-stage amplifier may be required to drive a device, for example a loudspeaker, which requires current to operate it. In this case the output stage of the amplifier must provide power instead of voltage amplification. There are two main differences between such stages and those used for voltage amplification. First, the valve must be constructed in such a manner that it allows a heavy current to flow: this in practice means larger valves. Secondly there is the difference in connecting the load: instead of having a load resistor it is usual to use a device, for example a transformer, which matches the output of the output stage to the device it is operating, to provide maximum power transfer.

OSCILLATORS

An oscillator circuit is one which has an output that varies in a periodic manner with time. This periodic variation may take a number of shapes, from simple sine waves to various pulse waveforms, occurring at frequencies which vary from a fraction of a cycle per second to thousands of millions of cycles per second. Oscillations below 25 cycles per second are generated best by electromechanical means but oscillations up to many millions of cycles per second are most efficiently produced by an electronic circuit. At frequencies in the thousands of millions of cycles per second range (gigacycles), ordinary valves will not operate so that special valves such as klystrons and magnetrons are used to generate the oscillations—much of the associated electronic circuitry of such valves being built into the valve itself.

Fig. 7 (c) shows the circuit of a simple oscillator. It can be seen that the input to its grid is a portion of its own output. If this is fed back in the correct phase (called positive feedback) the valve will continuously oscillate. An oscillator is in effect an amplifier which does not need an input—it supplies its own input. In the circuit the anode load is a coil of wire, that is an inductance, and across this coil is connected a capacitor, see Fig. 7 (a). If a range of sinewaves at various frequencies is fed into the valve the anode load circuit will behave differently at each of them. If the frequency is low, the coil behaves as though it were a straight piece of wire; the capacitor is thus shorted out and no voltage is developed across the combined anode load. If the frequency is high the capacitance will now behave as a short circuit and short out the coil which, on its own, would have had some impedance. In between these two extremes there is a frequency at which the inductance and capacitance do not short one another but instead combine to produce a resistance of high value, as shown in Fig. 7 (b). Across this imaginary resistance is developed an output voltage just as in the case of an amplifier. The

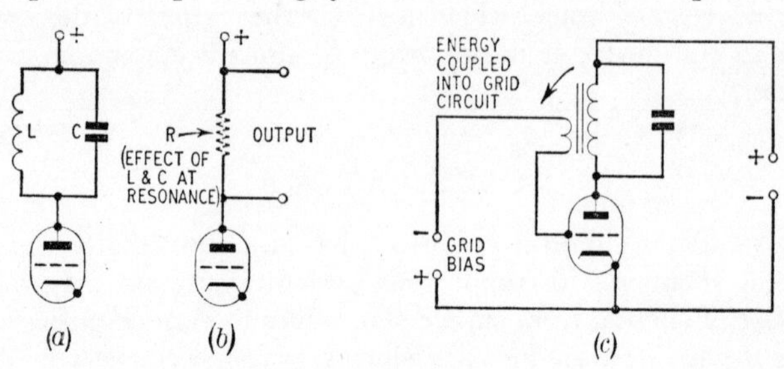

Fig. 7. (a) Triode valve with a resonant circuit connected to its anode. (b) At the resonant frequency, C and L comprise an equivalent resistance. (c) Simple oscillator circuit: some of the energy in the anode circuit is fed back to the grid of the valve.

frequency at which this effect occurs is called the *resonant frequency*, and is the only frequency at which the oscillator will operate. If some of the output voltage is now tapped off and fed back (in correct phase) to the input, as shown in Fig. 7 (c), oscillations will be maintained and the circuit will provide a continuous source of sine waves at a given amplitude and frequency.

There are many other oscillator arrangements. In all, however, the basic principle is that of positive feedback via a suitable network of 'frequency conscious' components between the electrodes of a valve or the sections of a transistor.

PRODUCING OTHER WAVEFORMS

The type of oscillator we have been discussing generates a sinusoidal waveform, a type of signal which is extremely useful throughout electronics practice. The other kinds of waveform, square wave, spike and sawtooth, have, however, to be generated by other techniques.

The square waveform

This type of signal can be produced from the sine wave produced by the oscillator just discussed. The sine wave should be at least ten times greater than the required peak value of the output square wave. This is necessary in order to obtain a square wave pulse the leading edge of which has a good rise time, that is, has a steep enough leading edge to give as near as possible a square shaped

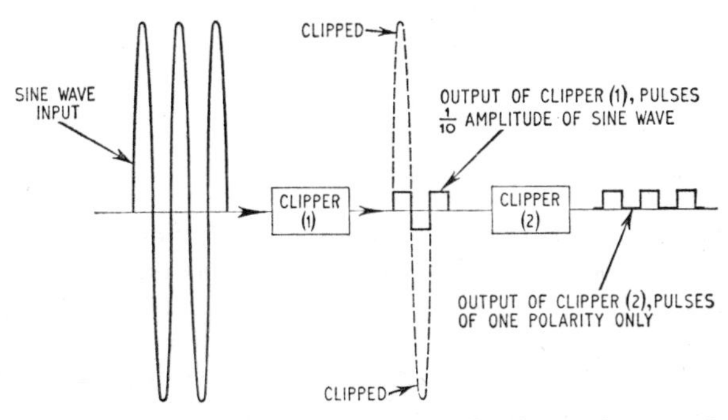

Fig. 8. Producing a square waveform by clipping a sine waveform.

pulse. To produce the square pulse the sine wave is clipped at a value equal to one tenth of the sine wave amplitude, as shown in Fig. 8. This produces nearly square pulses at a repetition frequency equal to the frequency of the original sine wave. If pulses of only one polarity are required, then the unwanted polarity can be removed by another clipper—clipper (2) in Fig. 8.

Clipper circuit

Fig. 9 shows a circuit using diodes to produce a square wave from a sine wave by clipping. The diodes may be thermionic diodes or semiconductor diodes—today they would most likely be the latter. Each diode is biased so that it will not conduct until the amplitude of the input signal has reached a prearranged value, which is of course the clipping value. At this value the diode conducts, short-circuiting the rest of the sine wave input to earth. The two diodes clip off the upper and lower portions of the sine wave above 10 per cent to produce a square pulse output. If only one

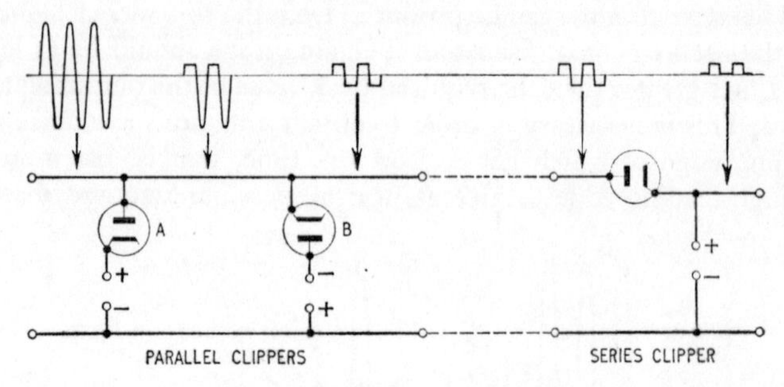

Fig. 9. Basic clipper circuits.

polarity square wave is required a series clipper diode may be used, as shown on the right of Fig. 9, to remove the pulses of the unwanted polarity. The series clipper will only conduct and pass pulses of the desired polarity, depending on how it is connected. In the diagram it is shown cutting off the negative part of the input to it, since it will only conduct during the positive part of the waveform.

Producing a spike waveform

One of the most useful waveforms is the spike shown in Fig. 10 (a) and produced by the circuit shown in (b).

Suppose that a square pulse is applied to the input, AB, of the circuit shown in Fig. 10 (b). Since the capacitor is uncharged it behaves as a short circuit so that it is possible to regard it as a conductor and to ignore its capacitive behaviour for a moment. At this stage therefore the full voltage of the input appears across the resistor. This state does not, however, continue for long because the capacitor begins to charge up. If it is of small value and if the resistor is also of low value (their product is called the *time constant* and is small in this example), the capacitor will charge rapidly and the voltage across the resistor will rapidly fall until the capacitor is fully charged and the voltage across the resistor is zero. This situation continues until the square wave input returns to zero. At this point we have a situation in which the source of voltage is in effect shorted out, leaving the voltage on the capacitor as the only voltage in the circuit. This represents a voltage potential difference across the resistor but of opposite polarity. The discharge current from

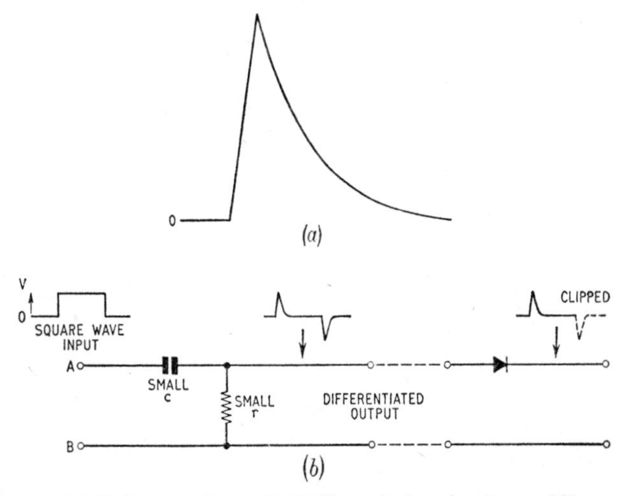

Fig. 10. (a) Spike waveform. (b) Differentiating circuit providing a series of spike pulses from a square wave input.

the capacitor gradually decreases, so that a tailing off occurs after the sharp leading edge—the same occurs when the capacitor charges. For one square pulse input, therefore, two spikes with sharp leading edges, have been produced and it is possible to select one of them, either positive-going or negative-going, by 'clipping' the output so that only the desired pulse, positive or negative, is passed on to some other circuit. This type of capacitance-resistance network with a time constant much smaller than the time for one pulse is called a *differentiating circuit*.

Producing a sawtooth

A further useful type of waveform is the sawtooth, which is used to produce the scanning on a cathode-ray tube and has many other applications including the variable triggering of other circuits, that is, if the input circuit of some control unit is set to fire at a predetermined point, it will do so when the sawtooth applied to the input reaches this predetermined value. In this way the unit can be

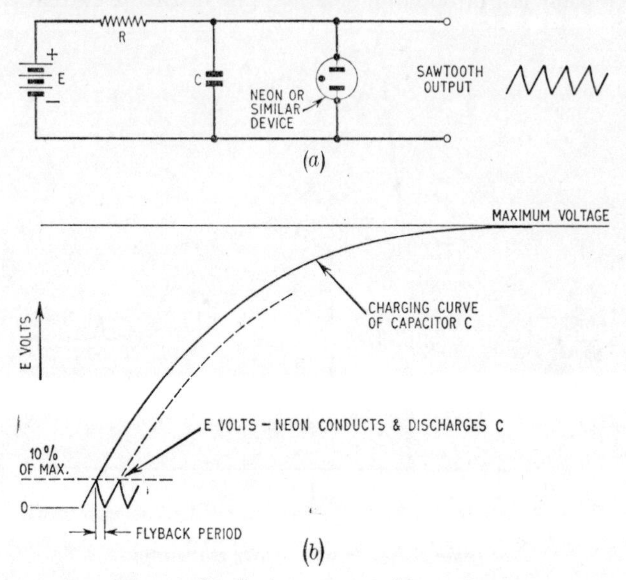

Fig. 11. Producing a sawtooth waveform.

triggered into activity over a range of time equal to the time taken for the sawtooth to rise from its reference value to its maximum.

A sawtooth waveform can be produced by allowing a capacitor to charge through a resistor, as shown in Fig. 11 (a), the capacitor subsequently being discharged in a very short time at some point along the charging curve. The time that the capacitor would take to reach its maximum charge is determined by the value of the capacitor, the value of the resistor and the magnitude of the voltage required to charge the capacitor fully. To obtain a linear slope to the sawtooth waveform all three should be made so that the voltage required to charge the capacitor fully is at least ten times greater than the desired amplitude of the sawtooth, see Fig. 11 (b). A neon tube, zener diode or some other 'electronic switch' device is arranged so that when the voltage on the capacitor reaches a certain value the device suddenly begins to pass a very high current so that the capacitor discharges in a relatively short time. This discharge time is called the flyback time and should be at least one hundredth of the time which would be required for the slope of the capacitor's charging curve to go from zero to its maximum voltage.

Pulse generators in general

These three methods of producing typical pulses are still frequently used. There are, however, many instances where a rise time of perhaps a few microseconds or even less is required of a square pulse, or an unusual mark-space ratio is required, or a very high repetition frequency, and so on and these special conditions

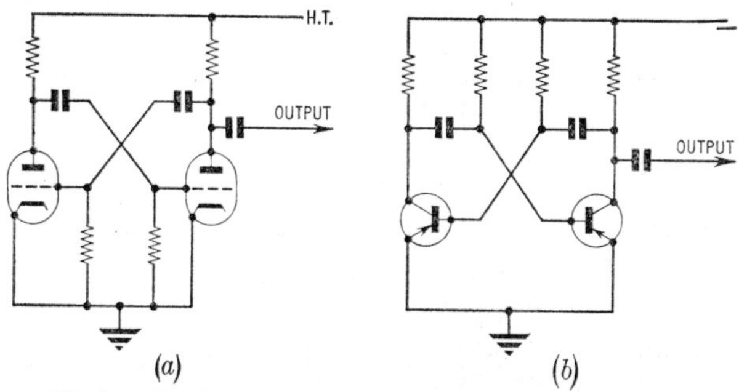

Fig. 12. Basic valve (a) and transistor (b) astable (free-running) multivibrator circuits.

demand more sophisticated circuits. The most important of these is the *multivibrator*, the basic form of which consists of two valves or transistors connected together as shown in Fig. 12. Each valve or transistor acts as a switch so that they switch each other on and off alternately producing, in combination with the resistances and capacitances in the circuit, a series of pulses. This circuit and others which perform in a similar manner can be grouped under three main headings, *astable*, *bistable* or *monostable*. In the multivibrator circuit the two valves or transistors are connected so that when one is conducting the other is silent, that is non-conducting. This is shown simply in Fig. 13.

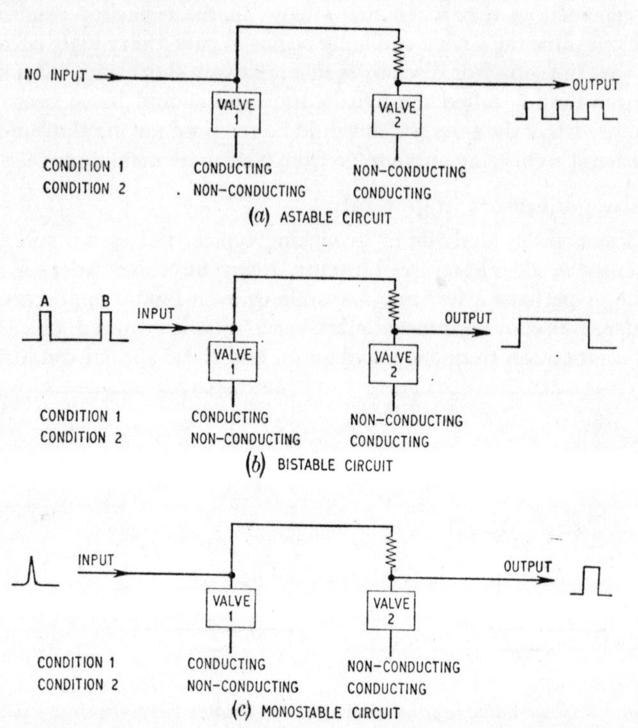

Fig. 13. (a) Astable circuit 'flip-flops' from condition 1 to condition 2 and back again and so on—that is, it is free-running. (b) The bistable circuit goes from condition 1 to condition 2 on receipt of pulse A, and remains there until the arrival of pulse B when it returns to condition 1 again. This gives one output pulse for two input pulses. (c) The monostable circuit is stable in condition 1. The spike pulse makes it go to condition 2, but it quickly returns to condition 1, its only stable condition. This arrangement gives a square wave output from a spike input.

Astable generators.—This generator operates freely at a predetermined frequency, the term astable meaning that it has no stable condition. That is, the change from one valve (or transistor) conducting and the other non-conducting to the opposite condition alters continuously. It is free-running and a common example is the free-running multivibrator.

Bistable circuits.—In this case both conditions are stable and it is necessary to apply a trigger pulse such as a spike to make one valve (or transistor) stop conducting and the other begin to conduct. When this happens the circuit will remain in the second condition until a second spike at the input makes it return to its original condition. The circuit is bistable, that is it is able to stay in either condition. This is very useful because as two input pulses are needed to complete one cycle of operation the arrangement can be used to divide by two, one output pulse being provided for every two input pulses.

The monostable circuit.—The monostable circuit is only stable in one of its two states, that is one valve (or transistor) conducting and the other cut off. An input pulse is needed to make the circuit flip over into the other condition and when it gets there it soon returns to its original stable state. This circuit is useful because using a spike input one can get a good square wave output with accurate dimensions.

D.C. RESTORATION

It is often necessary to restore the reference level of a signal. For example a square pulse might start from zero volts, its reference level, and rise to 10 volts, falling back to zero volts and so on. If this pulse is passed through a circuit which has a series capacitor, the reference level may be lost. The pulse may then, while still keeping its shape, start from zero, rise to 5 volts, fall ten volts to −5 volts then return to zero and so on. What has happened is that the reference level has been lost in passing the pulse through the capacitor. This is due to the standing charge which under certain circumstances may remain on the capacitor.

To restore the reference level a simple correcting circuit called a d.c. restorer circuit may be employed to put the square wave back into its original condition. The simplest type of d.c. restorer is shown in Fig. 14. It uses a single diode which may be either thermionic or a semiconductor. The capacitor is one of large value and the load resistor is also large so that the whole of the pulse is passed through the system without any distortion and if the diode was not there the capacitor would never acquire any appreciable charge. Consider what happens as soon as the pulses begin. The first negative excursion will cause the diode to conduct and the capacitor to receive a charge almost immediately because of the

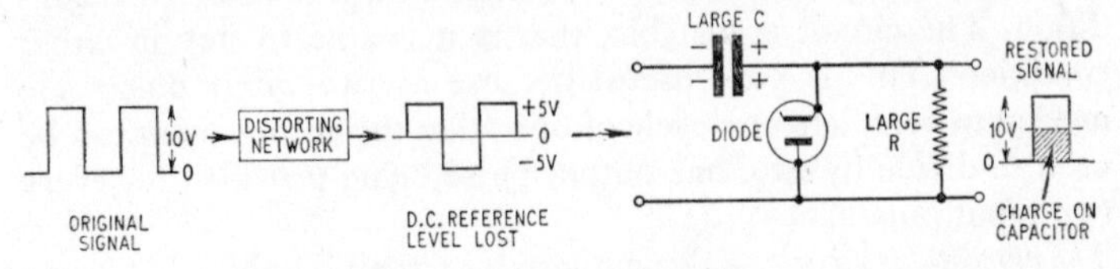

Fig. 14. D.C. restoration.

low resistance of the diode path when the diode conducts. This charge on the capacitor has the polarity shown in Fig. 14. The diode is then biased off by this voltage on the capacitor and will not conduct any more. The bias voltage cannot leak away through the resistor because this path has too high a resistance. The capacitor therefore retains the charge which is now added to the incoming signal every time and so restores the d.c. level. This is a simple system of d.c. restoration but it forms the basis of more complicated systems some of which allow a little of the charge to leak away during the interval between successive pulses so that, by arranging the values of the components in relation to the mark-space ratio, pulses can be restored to almost any level.

FEEDBACK

If an ordinary amplifier has a little of its output fed back so that it opposes the input signal it is said to employ negative feedback (as opposed to the positive feedback used in oscillators). Negative feedback has the effect of improving the amplifier's behaviour so much that it is regularly used even though it reduces the overall gain of the amplifier, sometimes by an appreciable amount. Feedback may be of current or voltage and whichever is used the effect on the input and output impedance of an amplifier is very marked. For instance if voltage feedback is used the output impedance of the amplifier is greatly reduced.

Cathode follower amplifier

This principle is used in the cathode follower circuit where the voltage feedback is such that the voltage gain of the amplifier is made less than unity—that is the signal output is less than the input. The output impedance is, however, very low indeed and this is a great advantage where it is necessary to feed power into a heavy load (e.g. a low resistance). The cathode follower is often found at the end of a chain of amplifying stages where its purpose is to match the last stage to the load.

The basic cathode follower circuit is shown in Fig. 15. The signal

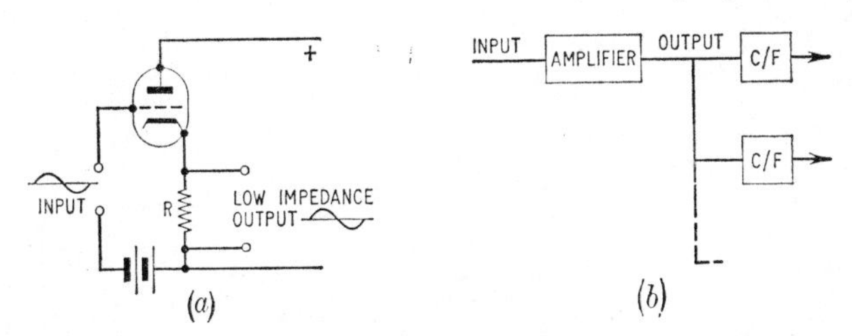

Fig. 15. The cathode follower. (a) Basic circuit, (b) a series of cathode followers used to feed a number of heavy loads from an amplifier.

is fed to the grid of the valve in the normal way but the output is taken from across the cathode resistor and there is no anode resistor.

This type of amplifier is used as a matching device to link an amplifier to its load, for example transmitters to aerials, radar receiver amplifiers to display systems, television studio output equipment to G.P.O. lines, and so on. It achieves this because its input impedance is high so that it does not 'load' the signal fed to it. Its output is a low impedance and can therefore feed into a heavy load.

5
TEST INSTRUMENTS

The electronics engineer needs a variety of test instruments with which to check equipment. Of all the instruments at the disposal of the engineer the moving-coil meter is perhaps the one most often used.

The moving-coil meter

If a coil carrying a current is placed in a magnetic field a force will act upon the coil. An instrument using this principle is shown in Fig. 1. The coil L is wound on a former of copper over a cylindrical iron core. The former is mounted on pivots so that it can

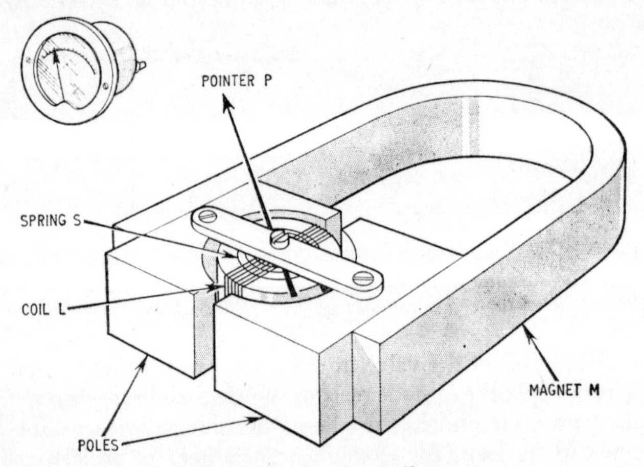

Fig. 1. The moving-coil meter.

rotate between the poles of the permanent magnet M. There is only a narrow air gap between the poles and the coil former, so that the magnetic field is radial and uniform for any position of the coil. The pointer, P, carried by the coil, is retained by the phosphor bronze spring S. When a current passes through the coil it tends to turn on its axis against the opposing force of the spring until it takes up an equilibrium position. This position is proportional to the current flowing through the coil.

The moving-coil ammeter can be obtained in many ranges, for example 0–50 microampere and 0–1 milliampere.

Measurement of current, voltage and resistance

The basic moving-coil ammeter arrangement that we have just described forms the basis of a number of other meters. By suitable modification it can be used to measure voltages and resistances. And its range as a current measuring meter can be greatly extended so that it will measure a number of different current ranges.

If a special low-resistance shunt is placed across the meter—see Fig. 2 (a)—a certain current—depending on the resistance value of the shunt—will flow through the shunt. The meter will then take only a proportion of the total current flowing through the whole arrangement. Thus a meter movement which is fully deflected when 1 mA flows through it can be made to deflect fully when 10 mA flows through the circuit, 9 mA going through the shunt. By arranging for shunts of various values to be placed across the meter, the meter will indicate currents of widely different magnitudes. Selection between different current ranges is easily effected by switching. The dial of the meter will usually have the same number of scales calibrated on it as there are ranges in the meter.

So far we have been considering the measurement of direct current. To measure alternating current a rectifying diode is connected in series with the meter to turn the a.c. into d.c.

To measure voltage, the meter is connected to the circuit under test via a series resistor—see Fig. 2 (b)—of appropriate value. A small current will flow through this series resistor, and as, in accordance with Ohm's law, this current is proportional to the voltage across the resistance, the scale of the meter can be calibrated in volts.

Including a battery in the meter enables it to measure resistance. Fig. 2 (c) shows the arrangement. The battery current flows

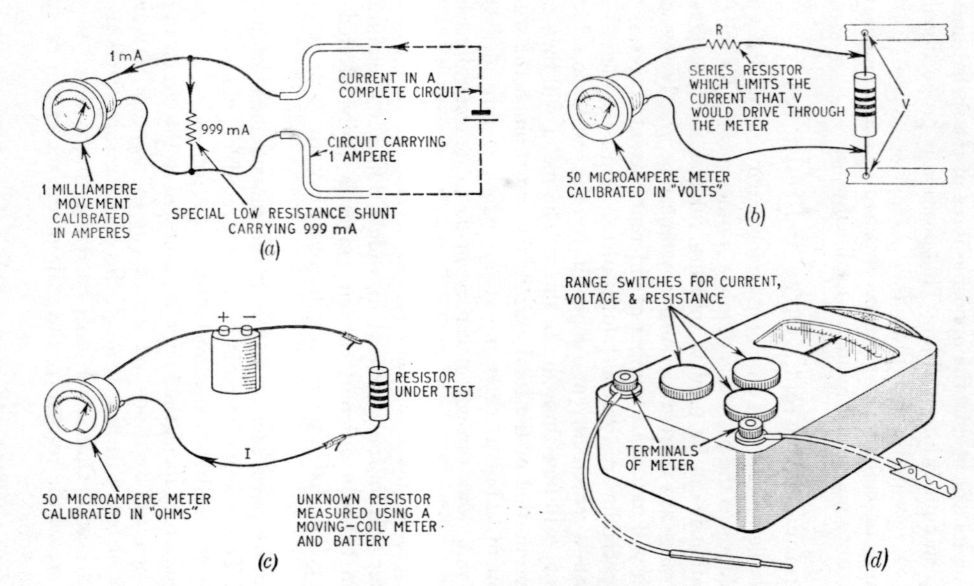

Fig. 2. Tests using a multimeter.

through the resistor under test and the meter: once more, following Ohm's law, the meter can be calibrated in ohms.

A meter incorporating these modifications is often called a multimeter.

The valve voltmeter

The multimeter can be used for most measurements but under certain conditions gives misleading results. For example, it may be suspected that the voltage on the grid of a valve is wrong. On measurement with the voltmeter section of the multimeter a very low reading is obtained. The person testing the circuit may immediately suspect a fault whereas in fact none exists. What has happened is that the voltage is reduced to a near zero level by the relatively heavy current drain of the meter. The resistance of the voltmeter is in fact lower than the source resistance of the circuit under test and thus the meter has short-circuited the grid of the valve.

This is illustrated in Fig. 3. The resistance R_s is at least ten times

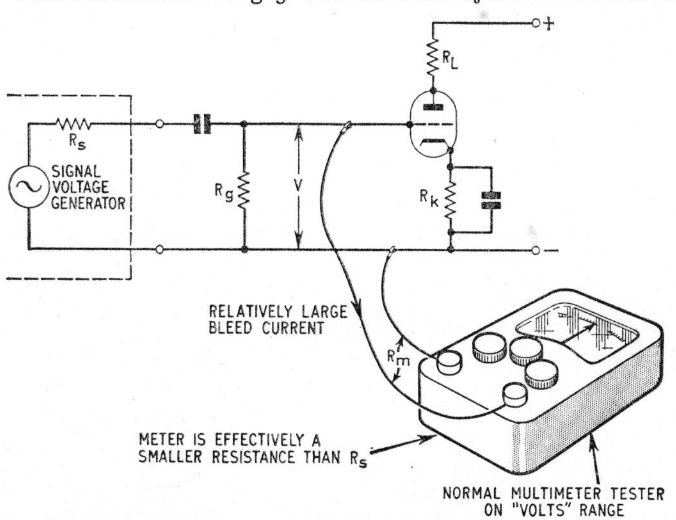

Fig. 3. The multimeter can give misleading results in some instances: its resistance alters the potential divider R_s–R_g.

as large as the resistance of the voltmeter section R_m of the multimeter. In the valve grid circuit under normal conditions this large value of resistance does not matter because the resistance of the valve grid circuit is even higher and therefore most of the signal voltage appears on the valve grid. However, when making a test the voltmeter resistance, which is relatively low—perhaps about 10,000 ohms—is placed across the grid circuit and the effective resistance of the grid circuit—which normally is several million ohms—is reduced to the low resistance value of the meter. As a result the voltage from the signal source is reduced by the 'potential divider' action of R_s and R_m, most of it being lost across resistance R_s. This reduction of voltage on the grid is not due to any fault in the circuit but is the result of connecting the meter across it. Under these circumstances the meter gives a low and erroneous reading.

There are many other circuits in electronics that reveal this basic deficiency of the simple multimeter. Because of this a more elaborate meter, the valve voltmeter, is needed. Its important charac-

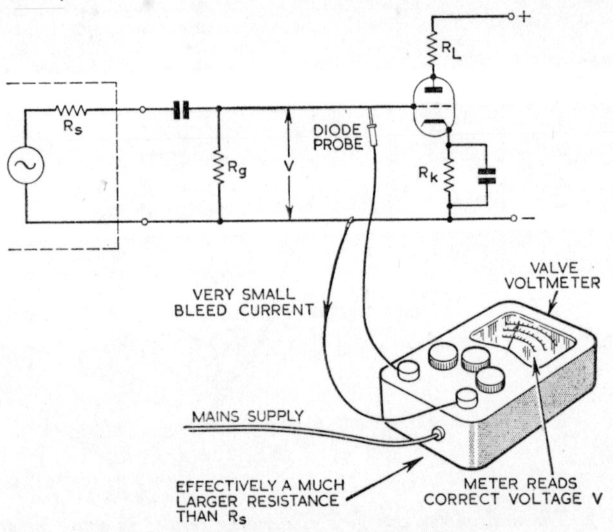

Fig. 4. The valve voltmeter and its use.

teristic is the high resistance it puts across the circuit under measurement. It may be considered as a very sensitive meter which draws an infinitesimally small amount of current when connected across the circuit. The simplest form of valve voltmeter uses only a single diode valve. More elaborate ones may use both triodes and diodes. Fig. 4 shows a valve voltmeter and its use in practice. Simple valve voltmeters measure only d.c. voltages, but the more expensive types can be switched to measure alternating voltages at frequencies from a few cycles per second up to many millions of cycles per second. This latter type has a special probe which rectifies the r.f. voltage before it is connected to the voltmeter to avoid loss in the lead cable which can occur at these higher frequencies.

The oscilloscope

One of the most useful 'tools' in electronics is the oscilloscope, the heart of which is a cathode-ray tube. A typical tube is shown in Fig. 5. To provide the necessary voltage to operate the tube, a long resistance chain may be used. This chain, which may have a total resistance of several megohms, has a voltage of two or three thousand volts applied across it called the e.h.t. supply (extra high tension).

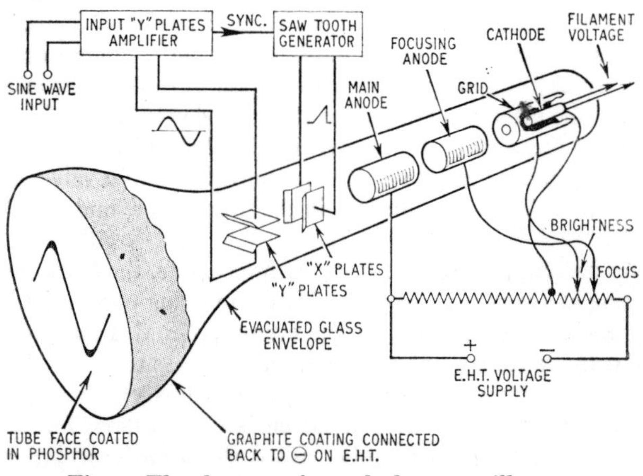

Fig. 5. The elements of a cathode-ray oscilloscope.

The cathode-ray tube consists of a specially-shaped glass tube evacuated of all gas, and containing a number of electrodes. The first of these is the filament which heats the cylindrical cathode to the correct working temperature. The electrons emitted by the cathode are attracted by the main anode and, as the anodes and the grid are in the form of cylinders, they pass through and eventually hit the screen at the end of the tube. The screen is covered by a phosphorescent material which glows when hit by the electrons, the number of electrons that hit the screen determining the brightness of the glow. The grid is connected to an appropriate place on the resistance chain and its potential relative to the cathode can be altered by varying this position. As the voltage on the grid determines the number of electrons that pass through it this voltage determines the brightness of the image on the screen. It is usually negative relative to the cathode.

After leaving the grid cylinder the electrons pass through a focusing cylinder. This is made negative with respect to the grid and tends to 'squeeze' the electrons into a pencil-thin beam, although the grid cylinder also helps in this process. Thus the result of these two electrodes is a sharp, finely focused beam of electrons. The focusing cylinder is also connected to the resistance chain and its position is made variable to give some control of focus. The electrons are now travelling towards the second cylinder, or main anode, which is highly positive with respect to the cathode and has attracted the electrons away from the cathode in exactly the same manner as the anode of a conventional valve. The grid cylinder acts very much the same way as the control grid of a valve except that in the cathode-ray tube it is in the form of a cylinder and not a mesh.

After passing through the main, or final anode, the electrons are travelling at a very high velocity. After impinging upon the face of the cathode-ray tube the electrons eventually find their way back to the cathode via a graphite coating on the glass tube.

One important difference between an ordinary valve and the cathode-ray tube is the value of the currents that flow. The anode current in the cathode-ray tube, called the beam current, rarely exceeds a few microamperes. Nevertheless on striking the end of the tube this current gives up enough energy to the phosphor coating to make it glow visibly: if the electron beam is stationary and not being deflected the tube will have a small bright spot stationary in the centre of the tube face. The colour of the spot may be green, blue, orange and so on, and it may have short persistence or long persistence, depending upon the phosphor used by the manufacturer and the final purpose of the tube. The term phosphor has nothing to do with the chemical phosphorus—it is used in this connection to describe any material used to coat the face of a cathode-ray tube because it glows when activated by electron bombardment. The term persistence is self-explanatory. A long persistence tube may glow for several seconds or more after bombardment. In measurement work in electronics short persistence tubes are used to prevent confusion of the beam movements, which may be changing over a period of milliseconds.

'X' and 'Y' plates

To produce a high velocity, properly focused electron beam is the function of the electrodes along the axis of the tube, but before the beam reaches the face it has to pass through two sets of deflector plates, the 'X' plates and the 'Y' plates. If a saw-tooth voltage is applied to the 'X' plates the beam will be drawn across the face of the tube and a visible line will be produced on the tube face. With the correct persistence value and the correct repetition frequency of sawtooth voltage the line appears to be permanent. When the sawtooth pulse reaches its maximum value it rapidly returns to zero and the spot will fly back to the beginning of its trace—so rapidly as to produce no glow at all. Therefore the important part of the sawtooth waveform is the relatively slow slope. The circuit producing the 'X' plate deflection voltage is called a timebase circuit.

The voltage to be observed and measured is applied to the 'Y' plates. If this voltage is a sine wave, i.e. alternating, it will appear on the face of the cathode-ray tube as shown in Fig. 5, the trace due to the sawtooth voltage on the 'X' plates being varied in a sinusoidal fashion by the voltage on the 'Y' plates.

Oscilloscope input

The input to the oscilloscope may be very small and in this case amplification is necessary before it can be used to produce a noticeable deflection on the tube face. The amplifier which does this is called the 'Y' plate amplifier, and a little of its output is fed to the 'X' plate sawtooth circuit for synchronizing the trace so that it begins at the appropriate time to allow the voltage on the 'Y' plates to appear correctly related to the 'X' plate scan.

Many types of commercial oscilloscope are available, from small transistorized portable instruments to large sensitive and complex laboratory models. In some applications it is necessary to display a wide range of frequencies, from nearly zero to many megacycles per second, on the 'Y' plates. Oscilloscopes that will do this usually have reduced sensitivity if, on the other hand, the range of frequencies is not so wide it is possible to obtain an oscilloscope which is very sensitive indeed.

Signal generators

Signal and pulse generators are two of the instruments commonly found in both the laboratory and workshop and since the advent of the transistor they can be made so light and portable that most field engineers carry one around as personal luggage.

A signal generator is a low-impedance source of a range of frequencies, its output being a fixed, or variable and calibrated, sinusoidal one. If the output is calibrated the generator is usually more expensive and is used mainly in laboratory work. The simpler and cheaper fixed-output instrument is used mainly for qualititive checks on circuits.

The term low impedance in this connection means that if the instrument is regarded as a generator of energy, which indeed it is, its internal impendance is low so that when current is drawn from it by the external load (the circuit under test) very little voltage is lost internally across the generator impedance. The calibrated-output generator must of course have a very low internal impedance in order that its output, once set, will not vary as the load current fluctuates during adjustments.

The cheaper type of signal generator is generally used, rather like a bell-buzzer, for continuity tests on circuits, to see whether signals are passing through equipment rather than to enable precise measurements of signals at particular points in the circuits to be made. This type of instrument is not beyond the pocket of the amateur. It must of course be accurate as to frequency—a stricture that applies to all kinds of generator.

Basically the signal generator consists of an oscillator which can be tuned over a series of appropriate ranges. The output at the chosen range is then fed via some circuit such as a cathode-follower to give a low impedance output. The range of frequencies covered may be from a few kilocycles to a few hundred megacycles. In many instruments it is possible to modulate the sinewave output—that is vary its amplitude in a regular fashion—by mixing it with a lower frequency sine wave. For example a one megacycle sine wave could have its amplitude varied between zero and a few volts at perhaps a 1,000 cycles per second (c/s).

When the frequency range required is in the low-frequency or audio range (10–20,000 c/s) it is difficult to obtain an oscillator with a high performance that can be tuned over the whole range. In this case a beat frequency oscillator (b.f.o.) system is used. In this system the outputs of two r.f. oscillators are applied to a non-linear impedance device such as a diode. This produces a range of output frequencies which includes the difference frequency between the two oscillators. It can be arranged so that this difference frequency, which is equal to the wanted audio frequency, is selected and fed to the output. The b.f.o. will operate over a range of frequencies from 25 c/s to several kilocycles.

Pulse generators

To test television equipment a video oscillator is used. This is a signal generator that gives a square wave output as well as sinusoidal output at frequencies ranging from a few cycles per second up to several megacycles per second. The output may be calibrated or not as in the case of the r.f. signal generator.

In addition to generators for video work there are many types of pulse generator available which offer various shaped pulses of many different lengths and having many different mark-space ratio settings. A typical output might be a square pulse with a rise time of 0·25 μsecs and a length of 5 msecs occurring at space intervals of 20 msecs—a pulse repetition rate of 40 pulses per second.

Bridges to measure inductance, capacitance and resistance

Other instruments used by the electronics engineer include bridge arrangements for the accurate measurement of resistance, inductance and capacitance. Bridges are circuit networks following the basic configuration of components shown in Fig. 6. Resistors only are used in the case of the resistance bridge (Fig. 6) but capacitors,

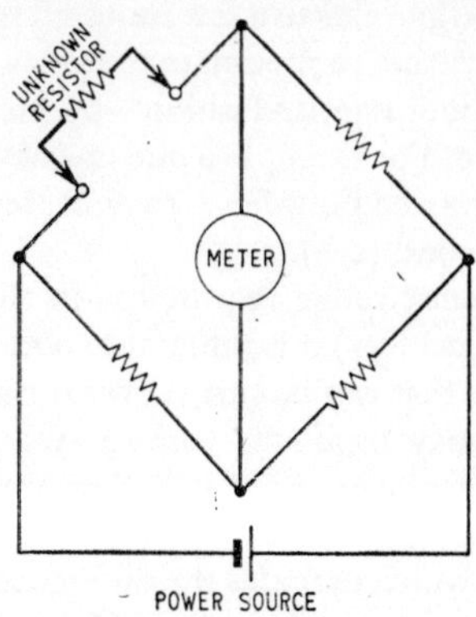

Fig. 6. Basic bridge measuring circuit.

inductors and resistors may be used in the cases of other bridges. The resistance bridge is based upon the Wheatstone bridge, and others being variations of the Schering bridge, Wien bridge and so on. The advantage of using a bridge circuit for finding the value of components is the greater accuracy and sensitivity of these circuits, facts which can be proved by mathematical analysis.

The bridge circuit is fed from some power source and incorporates some detector, for example a sensitive moving-coil meter. Both the power source and the detector are connected across different sections of the bridge and the unknown component is inserted in one of the four arms. The action of the Wheatstone bridge is described in Chapter 10.

Fig. 7 shows a typical commercial bridge capable of measuring all three quantities and so called a universal bridge. To do this it incorporates three separate bridges which are switched into the circuit at appropriate ranges. The power to the bridge is supplied

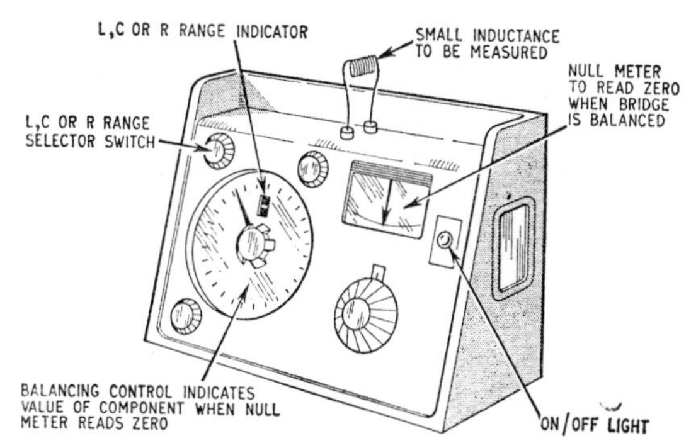

Fig. 7. Typical commercial universal bridge.

by an oscillator which can operate at either 1 kc/s or 10 kc/s. The bridge measures resistance from 0·1 ohms to 10 megohms, inductance from 1 microhenry to 100 henries and capacitance from 1 picofarad to 100 microfarads.

6

RADAR

The present advanced state of electronics is largely due to the impetus received from the development of radar during the last world war. The idea of radar was first thought of in the 1930s by a group of British scientists doing work on the propagation of radio waves. It was discovered during experiments that radio interference was caused by aircraft flying over the research station and on further investigation it was found that this was caused by radio waves being bounced off the aircraft and returned along the original path of transmission to the receivers on the ground. Development of this principle produced the first early warning radar equipment for detecting the presence of enemy aircraft.

In the early days radar systems operated at frequencies of the order of 50 megacycles per second and conventional circuitry was used, but subsequently most radar equipment was designed to operate at a frequency of about 10,000 megacycles, in the so-called microwave region. This was made possible by the development of the magnetron, an oscillator valve which can produce high peak powers at this frequency. The microwave signals are fed to and taken from the aerial by means of waveguides instead of the transmission lines used in normal radio practice. The use of microwaves simplifies the aerial system, making it possible to design aerials that gave a sharper, more clearly defined beam which aids in discriminating between near-by targets.

The principle of radar

The oscillator which is the heart of a radar set is pulse controlled so that it oscillates only for a very short period, often about one microsecond. These bursts of oscillation are used as the radar pulses. The pulses may be produced at intervals of one millisecond, that is a thousand pulses per second, and may be of relatively large peak powers—perhaps several megawatts in modern equipment—although the average power is low, perhaps only a few kilowatts.

The pulses of energy are fed to a highly directional aerial system and are transmitted along a path which may be almost as narrow as a searchlight beam. Immediately before the pulse is transmitted by the aerial the display system (which displays the information obtained by the radar installation), a cathode-ray tube begins an 'X' trace (see previous chapter). A fraction of the transmission pulse is fed to the receiver and appears on the cathode-ray tube as a small 'Y' deflection. This is called the *ground wave* and acts as a marker or reference for gauging the range of the echo or returned energy.

The transmitted pulse travels out into space and if it encounters a target, for example an aircraft or missile, or even in some instances a rain cloud, some of the energy is reflected back along the original

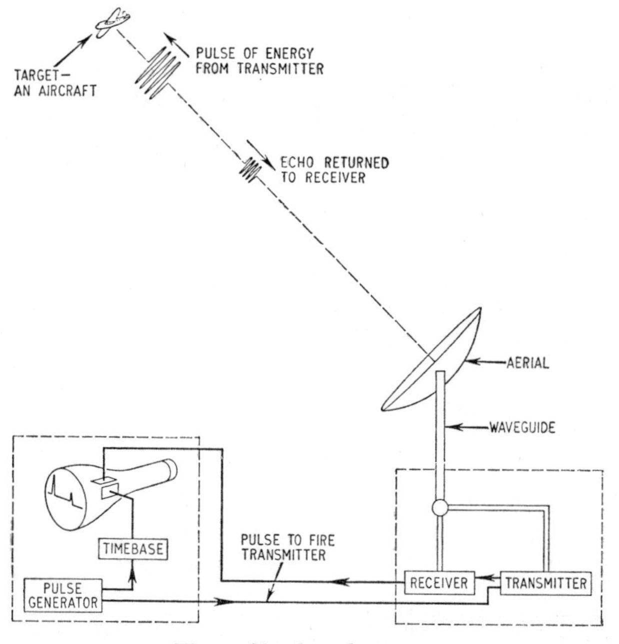

Fig. 1. Simple radar system.

path and is received by the aerial system. From the aerial it is fed into the receiving system and after detection and amplification is applied to the 'Y' plates of the cathode-ray tube, lifting the 'X' trace as did the ground wave. As the speed of electromagnetic energy in free space is known, the distance of the target from the aerial system can be calculated. And as the speed of the display tube 'X' trace is also known the screen of the display cathode-ray tube can be calibrated in miles or metres. The distance of the target from the radar set can thus be read off directly. As the energy has to travel to the target and back the distance travelled must be divided in half to get the true range when calibrating the range scale. Fig. 1 shows a simple radar system.

The radar transmitter

The part of a radar system which differs mostly from conventional radio systems is the transmitter. Fig. 2 shows a radar transmitter in simplified form. The magnetron oscillator is placed between the poles of a large magnet (needed for magnetron operation) and coupled to a waveguide which feeds the oscillator r.f. energy via a transmit or receive (T/R) switch to the aerial. The T/R

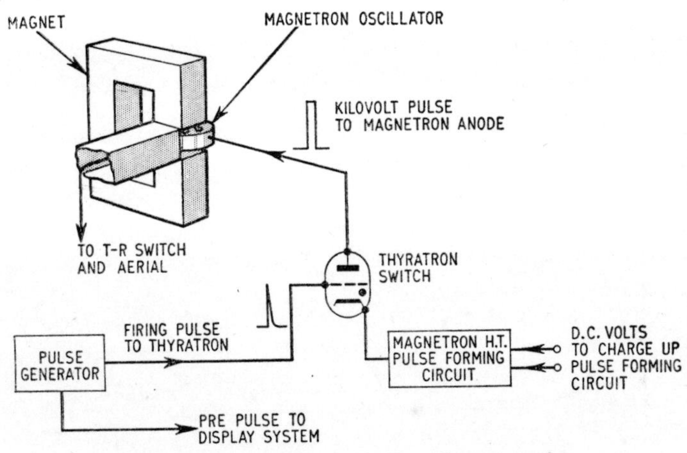

Fig. 2. The basic elements of a radar transmitter.

switch is necessary to prevent the powerful radar pulses that are being transmitted reaching and damaging the receiver (the returning echo pulse is only a minute fraction of the power of the pulse originally transmitted).

The magnetron is triggered into oscillation—'fired'—for a microsecond or so by a large positive voltage pulse applied to its anode. To ensure that the pulse which fires the magnetron is of the correct size and shape a special pulse-forming network charged from a d.c. voltage supply is used. The charging takes place in the relatively longer intervals between the magnetron output pulses, and it is therefore possible to obtain a very large voltage pulse to fire the magnetron from a relatively smaller d.c. supply.

The firing pulse is passed from the forming network to the magnetron via a thyratron valve used as an electronic switch. This valve, the operation of which is described in Chapter 3, is controlled by a small, accurate pulse provided by a pulse generator. The same generator may also supply a pulse to trigger the timebase circuits of the display system, thus ensuring that the display tube deflections and the radar pulse transmissions are synchronized.

The magnetron

The magnetron valve is the preferred source of high power oscillations in the microwave region. Recent years have seen many improvements in magnetron design but even now they are limited to mean powers of one or two kilowatts although the peak powers are measured in megawatts, enabling radar equipment to have a range of several hundred miles.

The action of the magnetron depends on the behaviour of electrons in crossed eleetric and magnetic fields. The d.c. electric field in the magnetron is supplied by the anode-cathode voltage of the thyratron during its conduction period. The magnetic field is constant, being supplied by a large permanent magnet situated, as shown in Fig. 3, outside the magnetron. When working on a magnetron care should be taken not to use heavy tools such as spanners and large screwdrivers made of ferrous materials: if they suddenly strike the body of the magnet, due to the attraction of the large magnetic field, they may damage the magnetron or distort the magnetic field. The engineer making any adjustment should also remove his wristwatch which may otherwise be permanently affected.

Under the influence of the magnetic and electric fields the electrons emitted by the cathode describe circular paths towards the

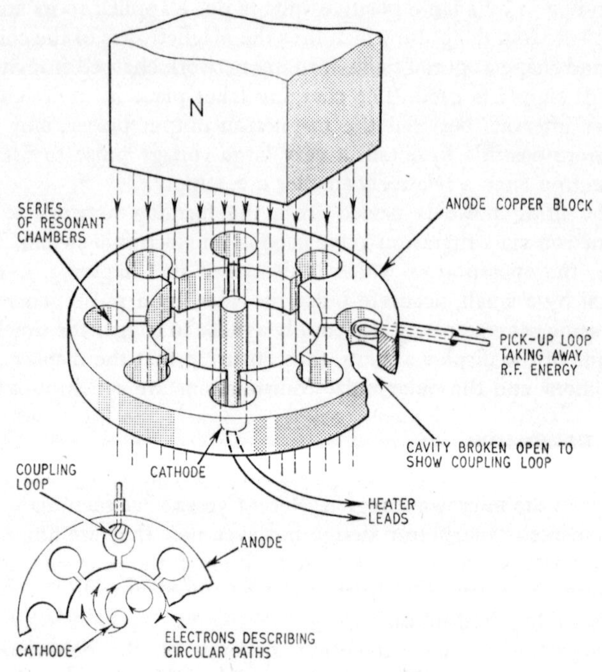

Fig. 3. The magnetron oscillator.

anode. Some of the electrons move in a tight circle and return to the cathode but others eventually reach the anode, which is a block of copper with a series of cavities cut into it, the mouths of the cavities, as shown in Fig. 3, facing inwards towards the cathode. Movement of the electrons across the mouths of these cavities sets up oscillations within the cavities and these oscillations in turn effect the movement of the electrons on their circular paths, producing a bunching effect. In this way energy exchanges occur between the cavities and the electron streams, the cavities being kept in oscillation during the application of the anode voltage pulse which provides the electric field. The frequency of the oscillations is determined by the physical size of the cavities—which in effect are tuned circuits and may be regarded as consisting of inductance and capacitance as illustrated in Fig. 4. In practice the cavities are strapped

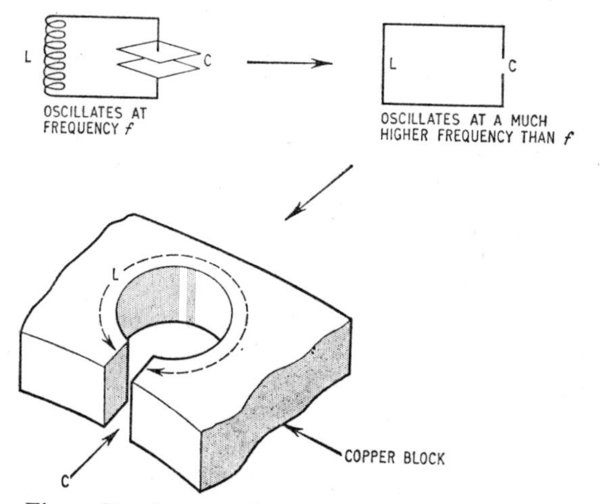

Fig. 4. Development of a cavity from an L, C circuit.

together in a particular arrangement so that the oscillations reinforce each other and the magnetron operates in its correct mode. Power is taken out of the magnetron to the waveguide by means of a loop of copper wire inserted in one of the cavities. This method of picking up r.f. energy, using a loop of wire or sometimes just a straight piece of wire sticking up in the r.f. field, is commonly used in microwave practice.

The thyratron valve

The thyratron valve has been already described in an earlier chapter. Its advantage as a control device lies in its ability to pass from a non-conducting state to a heavily conducting state in less than one microsecond. During the conducting period the resistance between its anode and its cathode is very low. When a switching rate of a thousand times a second is required hydrogen is used inside the thryratron instead of the more usual mercury vapour. The thyratron thus acts as a very high-speed switch.

During conduction, the large thyratrons used in modern radar can pass over two thousand amperes. The grid is the electrode used to make the thyratron conduct. To do this a sharp voltage spike (see Chapter 3 for description of a spike waveform) is applied to it. This voltage spike which fires the thyratron takes negligible power from the pulse generator. The thyratron is quenched—made to stop conducting—not by the grid, which loses control once the valve has started conducting, but by the reduction of the anode-to-cathode potential when the magnetron firing voltage pulse passes to the magnetron.

The pulse generator

The spike pulse to fire the thyratron may be provided by a free-running, astable multivibrator which will produce the correct shape of pulse for the thyratron grid. The pulse from the multivibrator is then amplified and shaped.

The pulse generator may also include cathode-follower outputs to other circuits such as the display equipment.

The transmit/receive (T/R) switch

If the transmitter pulse was allowed to pass through the whole aerial system uncontrolled it would also enter the receiver since in radar both the transmitter and receiver share the same aerial and waveguide system. To prevent this happening a T/R (transmit or receive) cell is inserted in the waveguide at an appropriate point. This cell consists of a tuned cavity in a glass envelope containing a gas which ionizes when the cell is 'struck'. The cell needs a 'keep-alive' voltage on a special electrode to keep the gas near to the ionizing point. When the transmitter is on, i.e. the magnetron emits a pulse, the cell is struck by this pulse and shorts out the part of the waveguide leading to the receiver, protecting it and allowing all the energy to pass up to the aerial.

At the end of the transmitter pulse the cell is extinguished and allows all the returned echo energy picked up by the aerial to pass to the receiver.

Recently a semiconductor T/R cell which does not require any keep-alive voltage and is of very small size has been produced: this may be the T/R switch of the future.

In practice the T/R switch allows a small fraction of the transmitted energy to enter the receiver to provide the ground wave pulse needed for reference purposes.

Waveguide system

In the early part of the nineteenth century scientists discovered that it was possible to send electromagnetic energy along a pipe of suitable dimensions. Although it is usual to think of an electric current as a number of electrons flowing along a wire, it is possible from a mathematical point of view to think of a magnetic and electric field configuration being guided along a path by a pair of wires. If this is accepted the idea of a waveguide does not seem so strange.

If certain conditions for the two fields, electric and magnetic, are satisfied it is possible to make them move along a guide of rectangular section. The conditions for a mode commonly used, the H_{10} mode, are that the electric field is perpendicular to the walls of the guide and the magnetic field is tangential to the walls, as shown in Fig. 5. Under these conditions copper walls may be fitted to the guide at specified distances apart so that the guide becomes a rectangular section of any desired length. If a probe energized by, say, a magnetron is inserted in one end of this arrangement then some distance away, perhaps thirty feet or more, almost the same amount of energy can be removed with another probe. This is proof that the energy has been successfully guided along the waveguide.

At very high frequencies (above 10,000 Mc/s) energy loss in a waveguide system is less than it is for any other method of transferring energy. Another advantage is that the mechanics of the aerial system are aided by the simple mechanical structure of the guide.

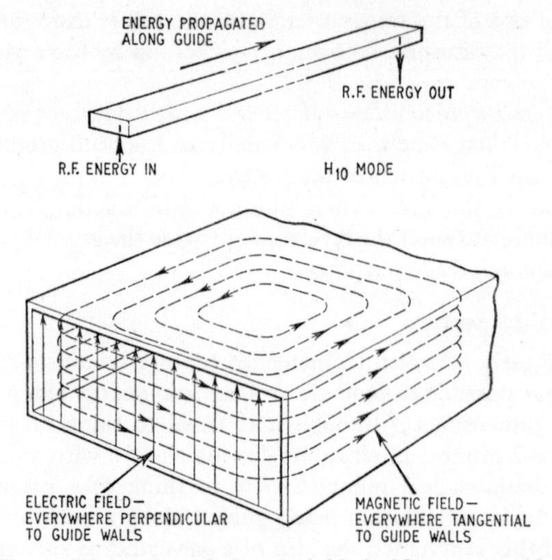

Fig. 5. Magnetic and electric fields inside a waveguide.

Aerials used in radar

In the microwave region of the frequency spectrum it is possible to use aerials which are small physically and which may have a variety of shapes. The shape of the energy beam that emerges from the aerial is determined by the shape of the aerial; many different kinds of aerial are in use according to the requirements of the system.

The effectiveness of an aerial is judged by its polar response, that is the extent to which it concentrates the transmitted energy in a beam. A diagram can be plotted to show this, such a diagram being called a polar diagram. This diagram shows the manner in which the field strength created by the transmitter varies with direction at a given distance from the transmitter's aerial. Fig. 6 shows a polar diagram of a cheese-dish aerial for two planes, vertical and horizontal. The wider the aerial dimension in any particular plane, the narrower the polar response in the plane. This type of aerial is widely used on ships for detecting other surface craft, say

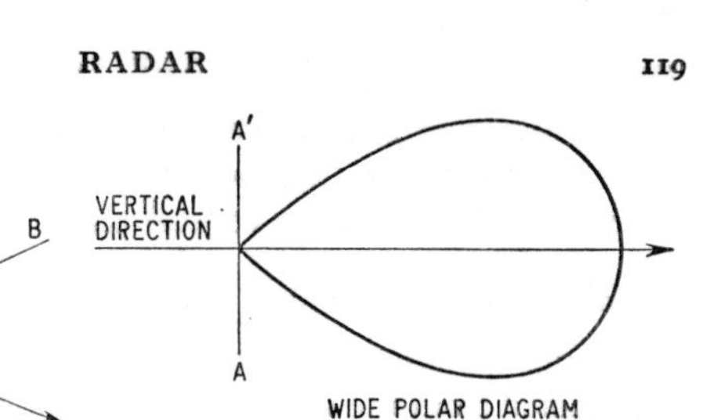

Fig. 6. Polar diagrams of a cheese aerial.

during a fog. The fine discrimination needed to determine accurately the position of other vessels is achieved by the narrow horizontal polar response, the wide vertical response being of no significance.

Fig. 7 shows a typical aerial used for the accurate determination

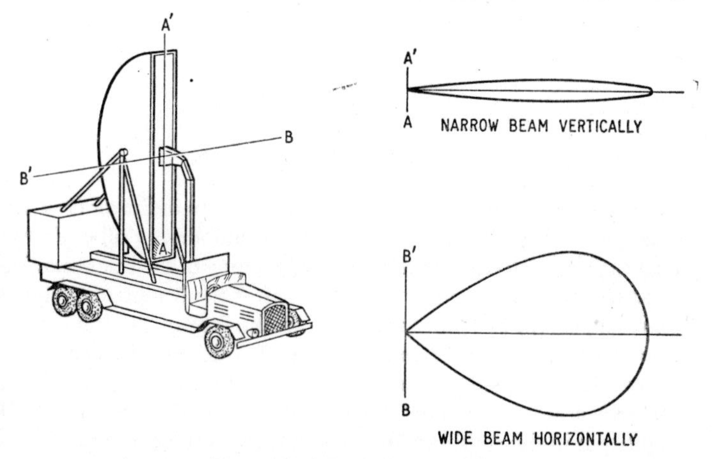

Fig. 7. Height-finding aerial.

of the elevation and hence the height of aircraft. This time the polar diagrams show that the wide part of the beam is horizontal and the narrow part is in the vertical plane.

Display of the signal

In Fig. 8 a typical 'A' scan display is shown. The ground wave is obtained from the initial transmitter pulse and used as a marker pulse. Returned energy after detection and amplification appears on the cathode-ray tube as a further 'blip'. Since the speed of radio waves in free space is known and the speed of the cathode-ray tube 'X' trace can be pre-set the real distance between the transmitter and the target can be inscribed directly on to the display tube face.

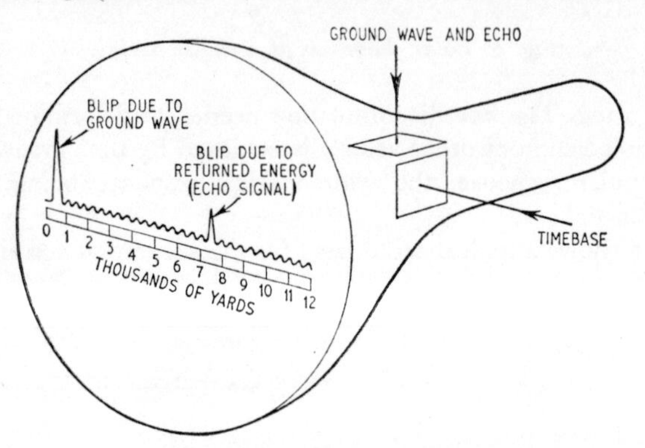

Fig. 8. Typical 'A' type scan.

Other types of display include the plan position indicator (P.P.I.). This type of display uses a special type of cathode-ray tube. Firstly it has high persistence—that is, any brightening of the trace remains for some time, perhaps many seconds, after the brightening signal has disappeared. Secondly the trace on the face of the tube starts from the centre and rotates in a series of widening circles instead of sweeping from side to side as in the 'A'-type scan. With P.P.I. if there is no echo signal nothing appears on the tube except a small bright spot in the centre to indicate the transmitter position. When an echo is received the scanning beam, which is rotating in synchronism with the aerial system, is suddenly brightened by applying the echo signal to the grid of the cathode-ray tube (instead of to the 'Y' plates as in the 'A'-scan). Thus not only the distance but the position of the target is directly shown on the display. The trace for a P.P.I. display tube is produced by special coils fitted around

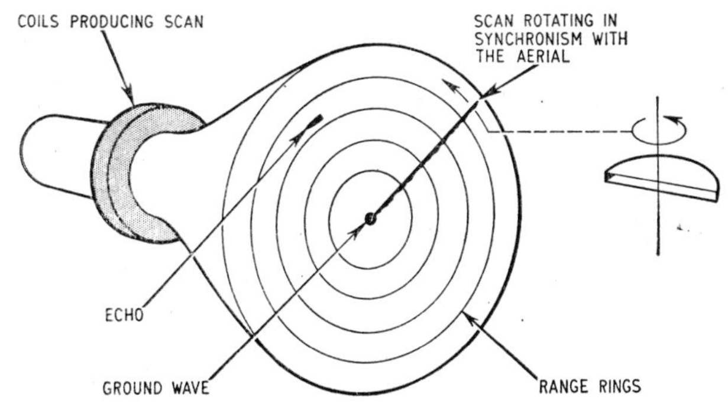

Fig. 9. A typical P.P.I. display.

the neck of the tube, as in a television receiver, there being no 'X' and 'Y' plates. A typical P.P.I. display is shown in Fig. 9.

Some more recent radar equipments use a display system which only shows moving targets. This arrangement is useful in harbours and similar situations where it indicates only something which is actually moving so that much useless information is excluded from the screen.

7
MEDICAL ELECTRONICS

Electronics today play an important role in medicine. This would be evident to anyone visiting one of the large teaching hospitals where up-to-date techniques in medicine and surgery are used and where large research projects are in hand. The co-operation between the medical profession and the engineering profession is unusually close and joint meetings are regularly held on electronic methods in medicine. In America university courses are held to teach the engineer something about medicine and the medical man something of electronics. Hardly a patient entering hospital for examination today will be exempt from the attention of some electronic apparatus.

What electronics can do in medicine

In radiography, the doctor uses a combination of X-ray and television techniques to enable him to examine a patient. This arrangement enables the patient to be examined for longer periods than with ordinary X-ray equipment as the amount of radiation required is much smaller than with conventional X-ray equipment.

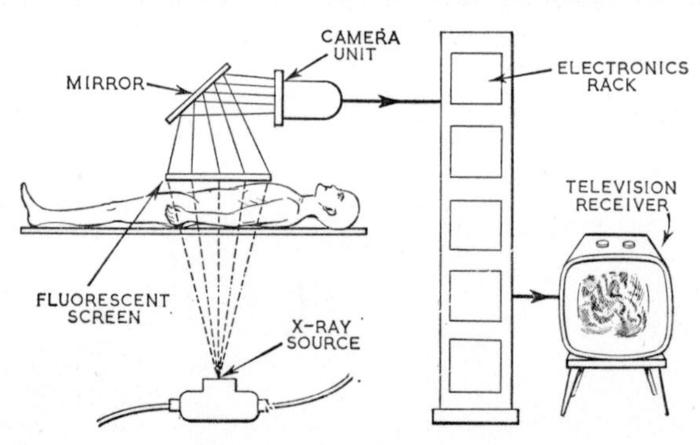

Fig. 1. X-ray image system.

This viewing can be done in normal daylight, which is a great advantage. Also the doctor can adjust the equipment so that it emphasizes different features of the X-ray, which aids his examination. The system is in effect a complete television system from camera tube right through to the receiver. Fig. 1 shows the X-ray image equipment in use to observe a patient. Radiation (X-rays) from the X-ray tube pass through the patient and produce an X-ray image on the fluorescent plate. This image is reflected by the mirror on to the face of the camera tube and is passed to the electronics rack which amplifies and processes it before feeding it to the television receiver.

Measuring human electricity

Electronic equipment is also used to measure the small physiological voltages generated by muscle and brain tissue. These voltages are of importance in diagnosing conditions of the heart and

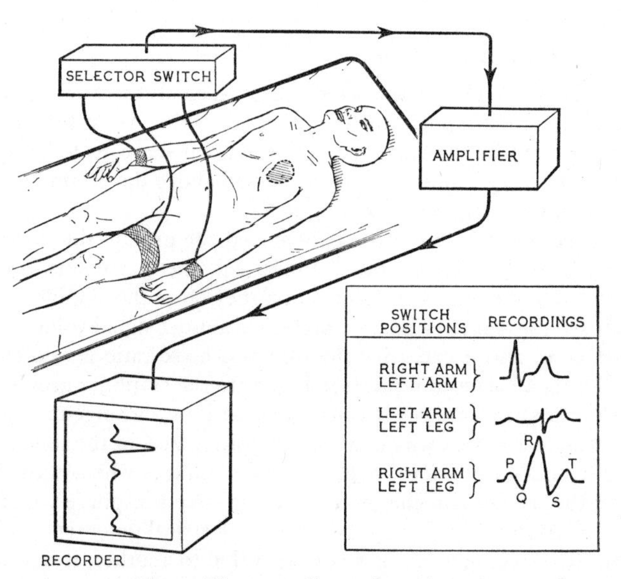

Fig. 2. Electrocardiograph equipment.

abnormalities of brain function. The heart is a powerful muscle and in functioning it generates voltages which radiate throughout the body. The doctor can measure and record these signals by means of electrodes connected to the patient's limbs. These voltages form a pattern that is the same in all normal people so that divergencies from this pattern indicate some form of abnormality.

Fig. 2 shows a typical recording and the measurement being made on a patient, the technique being called electrocardiography (e.c.g.). The signals generated by the heart are picked up by electrodes attached to the wrists and the left leg of the subject. Measurement is made in turn of the variations in potential at the right and left arm, right arm and left leg and finally left arm and left leg. The voltages picked up in this way are fed to a suitable amplifier called a physiological amplifier. The gain, impedance and bandwidth requirements of these amplifiers vary slightly with different applications but one thing that they all have in common is a balanced input.

Physiological amplifiers

In Fig. 2 the electrodes, which may be small discs of silver or platinum, are strapped to the patient's skin. In position the electrodes can be thought of, in the electrical sense, as being connected to a large conducting mass situated above the ground which is another even larger conducting mass.

Voltages exist between the patient and the ground due to stray magnetic fields from various sources, e.g. the a.c. mains. These voltages are of course unwanted and if they enter the amplifier they will mask the wanted signal voltages. Unwanted signal voltages are called *noise*, and if excessive the signal-to-noise ratio is said to be low. Using a balanced input (see Fig. 3) to the amplifier avoids the pick up of stray voltages because such an input arrangement only responds to voltages which are in opposite phase in the leads, the so-called out-of-phase voltages, rejecting voltages which occur between the patient and the ground, the so-called in-phase voltages. Stray interference is usually a change in electrical potential—either from zero to some value or from one value to another, it does not matter which. What does matter is that these changes are induced into any piece of wire which is adjacent to the source of this noise. When the wire happens to be connected to the input of a valve amplifier, the noise is amplified and appears at the valve's output together with the wanted signals. If a balanced input arrangement (Fig. 3) is used, employing two valves as shown, both grids will be similarly affected by the interfering voltage, and by its design the amplifier can be arranged so that it cancels out the two in-phase

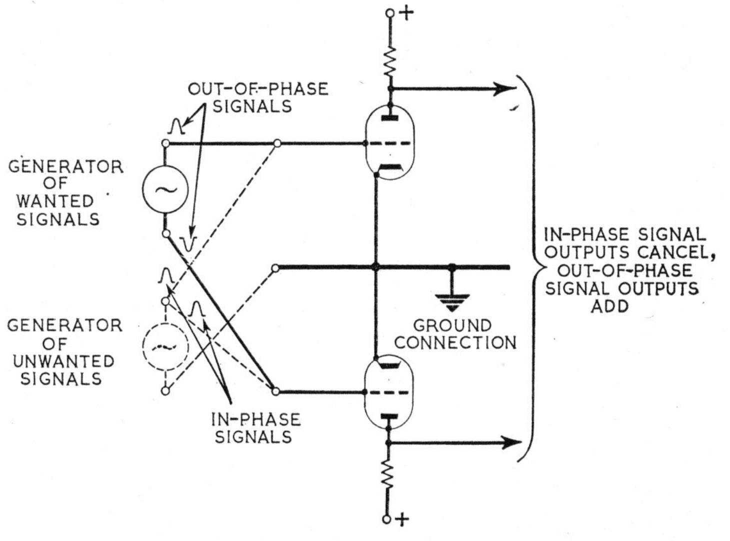

Fig. 3. Principle of the balanced input stage.

inputs to the first stage grids. This amplifier design is important because of necessity the amplifier must be attached somewhere to the ground, either directly to obey safety regulations or indirectly through its own power supply.

With the use of transistors the problem is simplified since a transistorized physiological amplifier need not be in contact with the ground and has no connection with the earthed mains supply. By using transistors it is possible to design amplifiers for use in medicine which have a very high signal-to-noise ratio. The only problem is that in the normal transistor amplifier configuration the input impedance is low: a high input impedance is required for most physiological amplifiers since the source impedance of the generator supplying the input—namely, the patient, has a relatively high impedance. To overcome this problem amplifiers are available which use a thermionic valve for the input stage and transistors for the later stages. Such amplifiers are called hybrid amplifiers.

The output of the physiological amplifier is applied to a pen recorder which makes a permanent record of the signals from the patient—an electrocardiograph. The usual type of pen recorder cuts off at about 100 c/s and as this is too low for the adequate presentation of e.c.g. recordings special recorders are used which operate successfully up to 1,000 c/s.

There are today many portable e.c.g. machines available which can easily be taken to patients in their homes for the examination of heart conditions.

Voltages from the brain

The investigation of brain functions can be undertaken with an equipment called the electroencephalograph (e.e.g.). The technique

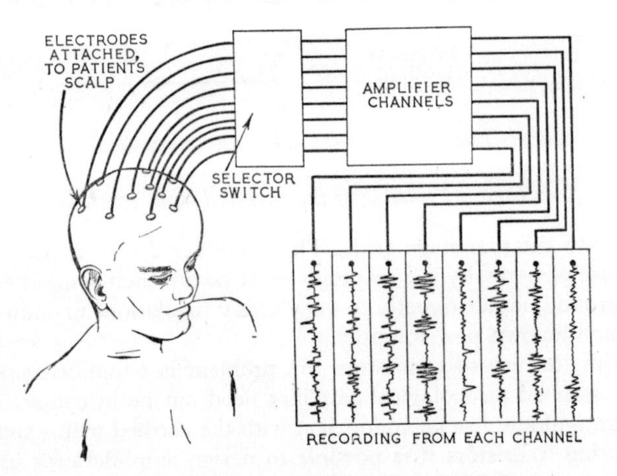

Fig. 4. Electroencephalograph equipment.

used is to measure the random voltages generated in the millions of nerve cells of the brain. The amplifiers used with this equipment need to have higher gains than those used in e.c.g. since the input signals from the brain may only be about 100 microvolts.

The voltages are detected at the patient's scalp by using small disc electrodes firmly attached to the skin by a harness. It is usual to use up to twenty-four electrodes. Each provides a series of signals which are fed to separate amplifiers and recorders. These amplifiers and recorders are built into a single complete unit as shown in Fig. 4. To save expense there are generally about sixteen amplifying channels and a switch is provided so that the physician can select various areas of the brain for investigation. The output of the selector is led to a pre-amplifier, one for each channel, and then to the main amplifiers and recorders.

Stimulators

The use of electronics to stimulate reactions in patients is perhaps most used in the field of cardiac medicine. The heart beats as the result of a nervous impulse which is conducted through the entire heart, beginning in the auricles and ending in the ventricles. This nervous impulse is electrical and can be imitated by electronic means. It should not be confused with potentials measured in e.c.g., the latter being the *result* of the heart beating. In some diseases a temporary or permanent disorder of the conducting mechanism occurs so that the heart beats irregularly. Stimulators may in these circumstances be applied either externally on the chest wall or internally to the actual heart muscle. The stimulator produces a spike or similar waveform at regular intervals to make the heart beat with a regular rhythm. In chronic disease of this type a small semiconductor stimulator receiver can be sewn into the actual heart muscle; externally the patient wears a transistor transmitter, powered from a small battery, and this supplies the necessary impulse for the internal receiving equipment to stimulate the heart to make it beat regularly. These instruments can be left in the patient for many years without attention.

A danger at the conclusion of a major heart operation is for the patient to suffer from a heart blockage due to lack of conduction between the auricles and ventricles. To overcome this electrodes can be sewn to the auricles to pick up the electrical signals they generate. These signals—a series of impulses—are passed to a delay circuit which delays them the exact amount of time they would be

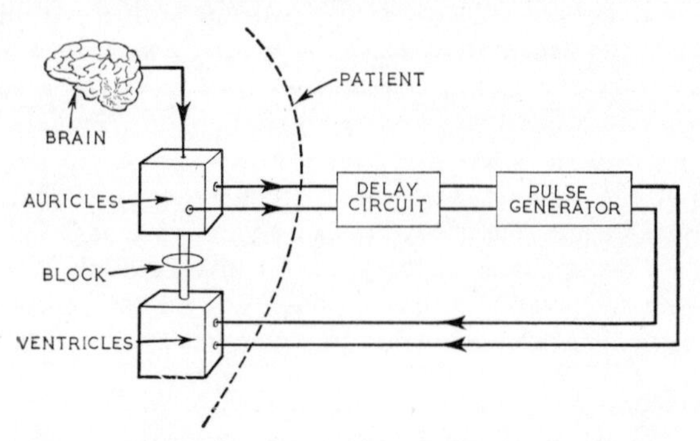

Fig. 5. Electronic means of correcting a heart block.

delayed during conduction from the auricle to ventricle in a normal person. After delay the impulse is used to trigger a pulse generator the output of which is used to stimulate the heart to beat through another set of leads attached to the ventricles. This arrangement is shown in Fig. 5.

Electronics helps the paralysed

Another use of this technique is to operate artificial muscles in paralysed patients. In some diseases the muscle is useless although the nervous conduction system which controls it is intact. The scientist endeavours to pick up the control signal from the brain to the muscle and uses this to operate an artificial muscle attached to the paralysed limb. Fig. 6 shows an experimental use of the idea. The artificial muscle consists of a rubber tube covered with a long-weave jacket of metal wires. When the tube is expanded by pumping carbon dioxide into it the metal jacket shortens in length, thus acting like a muscle. In this application the artificial muscle is attached to one of the fingers and the appropriate nervous control impulse is obtained at the forearm. This impulse is taken to an amplifier whose

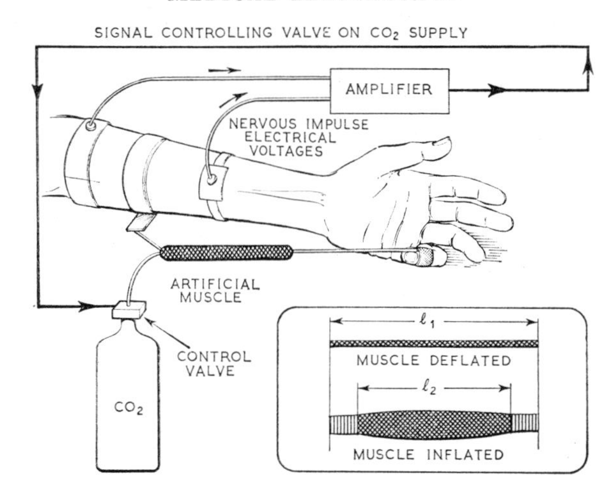

Fig. 6. An artificial muscle which can be electronically controlled.

output is used to control the mechanical valve operating the CO_2 supply which expands the artificial muscle. This system was first reported in the proceedings of the Second International Conference on Medical Electronics.

Inside the stomach—the radio pill

Radio pills are, as their name suggests, devices to be swallowed. The idea is that they transmit information which can be detected, recorded and later assessed by the physician during their passage through the human alimentary system. Before the advent of the transistor it was impossible to make a transmitter small enough to be swallowed without considerable discomfort to the patient. Now, by using a transistor circuit powered by a suitable cell, pills can be made which are less than an inch in length and half an inch in diameter. Fig. 7 shows a typical radio pill which will transmit information about the pressure, temperature, acidity, and many other factors of interest to the physician. The pill is essentially a minute radio transmitter working at about 3–400 kc/s and the information is radiated through the body to a sensitive receiver outside which detects the signals and passes them to a recorder. Information is conveyed by varying the frequency of the transmitted signals (frequency modulation), and this can be done in several ways. If for example it is desired to measure the pressure inside the intestines, then the inductor in the circuit can be made to alter the tuning of the transmitter, its iron core being displaced in accordance with the pressure on the walls of the pill. If the acidity of the stomach is to be measured this can be done by placing pH electrodes on the surface of the pill. When immersed in the acid

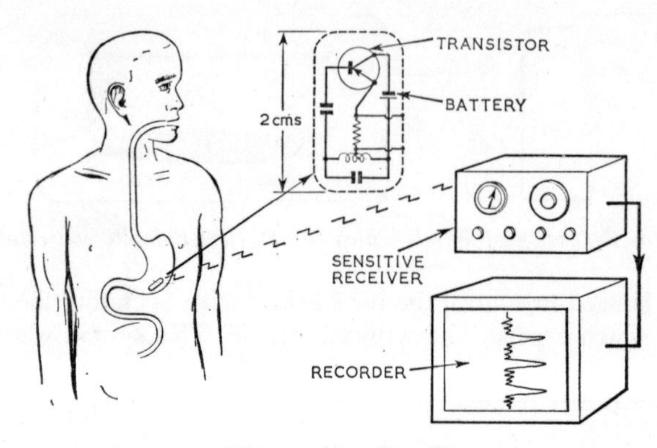

Fig. 7. A 'radio pill'.

solution in the stomach the electrodes generate a voltage proportional to the acidity, and this is used to change the frequency of the transmitter by altering the potential on the transistor. Again if the temperature is to be measured then the effect of temperature on the transistor can be used to vary the frequency.

Although the pill-transmitter can hardly be an accurate, close-tolerance device the receiver should be as selective as possible allowing for the bandwidth required to receive the frequency changes from the radio pill. Typical radio pill transmitters may drift by as much as one per cent off their nominal centre frequency so that the signal frequency variations should be at least ten per cent greater to give a good signal-to-noise ratio.

Measuring blood pressure

Electronic equipments are available to monitor continuously a patient's blood pressure. This is done using a finger cuff which is inflated to a variable extent periodically. At the same time the pulse beats in the finger are recorded. When the recording chart ceases

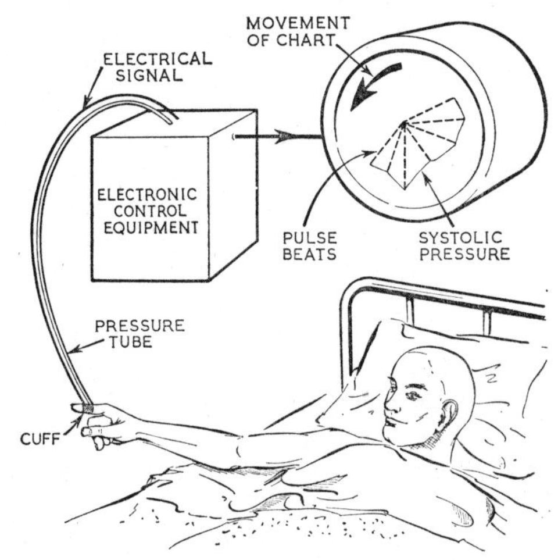

Fig. 8. Automatic system of monitoring blood pressure.

to record pulse beats the systolic blood pressure and the pressure to which the finger cuff is inflated are equal. The information is recorded on a very slowly moving circular chart as shown in Fig. 8. Although the equipment is based on general engineering principles the recording and control devices are all electronic.

Cancer research

In the field of medical research the electron spin resonator is an important instrument, especially in connection with cancer research.

The development of the technique of electron spin resonance (e.s.r.) was largely the result of radar research during the last war. In fact for many years after the war most of the experimental e.s.r. machines were made of government surplus equipment. In e.s.r.

apparatus microwave equipment, including waveguides, is used and thus many of the techniques used in military radar equipments are also applicable to e.s.r. medical research equipment.

The burnt surface of a charred slice of toast consist of molecules that have been damaged by excess heat. These molecules, called free radicals, consist of a group of atoms imperfectly bonded together and with an unpaired electron rotating around them in such a way as to act as though the molecule has an electric current flowing in a circle around it. This current produces a magnetic field which, together with the spin of the unpaired electron, makes the molecule behave like a spinning magnet. If the free radical is placed in a large magnetic field it acts like a spinning top under the influence of gravity. It begins to precess—that is it continues spinning but at the same time describes a slow circle around the axis of the large magnetic field. If the radical under these conditions is exposed to a radio frequency field at a certain frequency it will absorb energy from the r.f. field. This is rather like trying to make the spinning top take an upright position again—energy is needed to oppose the pull from gravity. In the same way the spinning radical will describe a smaller circle in its precession after absorbing energy from the r.f. field.

This absorption of energy can be detected and recorded. This recording may take the form of a curve, and from the size and the shape of the absorption curve quite a lot can be deduced about the number and type of radicals present. The role of free radicals in producing cancer is not yet fully understood but their presence is thought to be significant: hence e.s.r. equipment is assisting in the research for a cure for cancer.

The radio frequency energy used is in the microwave region, as in radar. Fig. 9 shows an elementary e.s.r. equipment. The substance under examination is placed, in a glass tube, in the waveguide between the poles of a large magnet. R.F. energy from a klystron valve (an oscillator valve which, like the magnetron, is capable of producing oscillations of the order of 10,000 megacycles per second) is fed into the waveguide at one end. The power required is small—perhaps only about 40 milliwatts—and the klystron oscillates continuously. If free radicals are present in the substance under investigation some of the microwave energy transmitted down the guide will be absorbed so that a reduced power reaches the end of the waveguide. The power at this end of the waveguide is detected by a crystal detector fed by a probe inserted in the waveguide, and the output of this is fed to a suitable amplifying system. Thus the amount of energy absorbed by the substance under investigation can be measured. The signal from the amplifier may for convenience be displayed on a cathode-ray tube.

There is only one frequency and one magnetic field strength for a substance under investigation that will satisfy the conditions re-

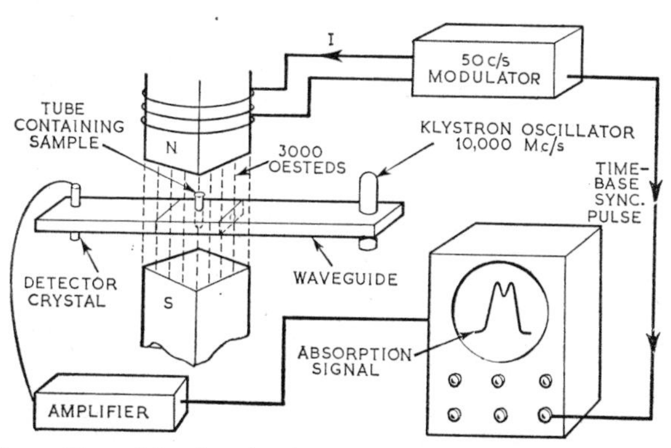

Fig. 9. Essentials of a simple electron spin resonance equipment.

quired for absorption to take place. In the case of burnt toast it is about 10,000 megacycles (r.f. energy) and three thousand oersteds (strength of magnetic field). If therefore the strength of the magnetic field is varied above and below this critical value, which can be done by passing a low frequency current (say 50 c/s) through a suitable coil wound on the magnet poles, then it is possible to obtain an absorption curve, rising as the magnetic field approaches the critical value and thereafter declining, which can be displayed as a trace on a cathode-ray tube in the same way as any other low frequency signal. Some of the current which modulates the magnetic field in this way is also fed to the cathode-ray tube timebase so that the X trace and the absorption signals are synchronized.

Electronic stethoscope

The electronic stethoscope consists of a microphone, mounted in a rubber cover, the output of which is connected to the input of an amplifier. The whole system is required to respond to signals between 0–10,000 c/s, with equal amplification over the whole of this band. The output of the amplifier is taken to a selection of filters which enable certain parts of the band to be selected and recorded. If a series of recordings is made with different filters in circuit,

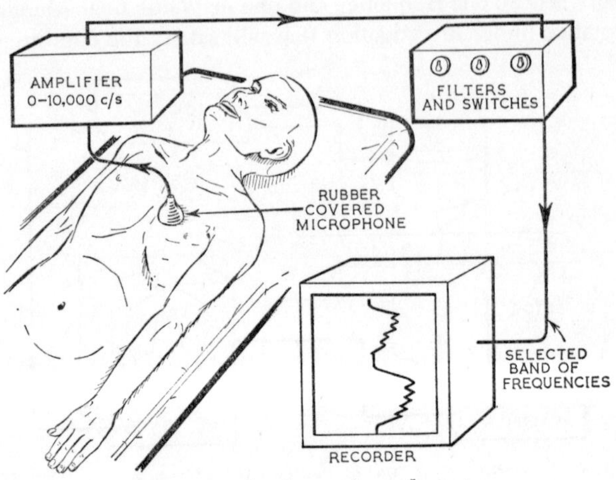

Fig. 10. An electronic stethoscope.

these can be compared with recordings made on a healthy person so that any unusual responses are revealed. From this information the doctor can deduce what damage to tissue might produce the effect. The same technique can be used to detect heart troubles. The fact that certain frequencies can be excluded by using the filters makes it possible to detect the presence of ominous sounds without them being masked by noises at other frequencies. An electronic stethoscope is shown in Fig. 10.

Ultrasonics

Ultrasonic techniques have a certain similarity to radar. Bursts of ultrasonic (i.e. frequencies above about 16,000 c/s) energy, pulsed as in radar, are transmitted and reflected and the returned echoes are received and can be plotted on a recorder. This energy can be applied at certain points to the body, for example the skull and the abdomen. The technique provides information which can be used as a corollary to X-ray examination and other techniques to tell the physician more about his patient to aid his diagnosis.

A further use of ultrasonics in medicine is the ultrasonic hypodermic needle. The needle incorporates a suitable transducer to which is applied the output of an ultrasonic generator. The ultrasonic transducer makes the needle act rather like a road drill. Its advantage is that it can be used for penetrating deep body tissues without damaging surrounding tissue or causing pain.

The sterilization of surgical instruments is another sphere where ultrasonic techniques are playing an important part. The ultrasonic transmitter (e.g. a piezo-electric transducer) is placed in the sterilizing tank together with agents, for example detergent, to lower the surface tension of the water. When the transmitter is switched on the micro-agitation it produces in the water removes dirt from the instruments and makes subsequent sterilization much easier.

8
ELECTRONICS IN SPACE

By electronic techniques information on the conditions in space can be obtained and sent back to earth. Instruments can be designed to operate under conditions which might seriously impair the judgement of a human being. And of course with manned space flight electronic equipment assists by making the rapid assessment and storage of a vast quantity of data possible. The need for electronic equipment in space vehicles as research programmes build up makes the problem of the design of suitable equipment a most important one. The microminiaturization of electronic components and the need to increase their reliability are two of the most vital problems in the field of space instrumentation.

Reliability of components

The behaviour of electronic components is a major problem in the environment of space. Apart from the need to make components as small and light as possible the designer of electronic components has to make sure that they are reliable and capable of working under arduous conditions, unlike any to be found in normal practice.

Some of the environmental conditions under which components have to work are outlined below.

Launching shock

This is due to the enormous acceleration that occurs when a missile is launched. There is great vibration during this period and this may continue throughout the missile's life. The vibration can displace relays, upsetting their working, and also makes leads vibrate, producing microphony in valves and introducing spurious noise signals into the equipment (which may already be operating at a very low signal level). These effects can be overcome to some extent by using potted circuits and printed wiring. The former technique can be used for an amplifier, switching circuit, multivibrator or any circuit which can be regarded as a complete unit or module. The circuit to be encapsulated in this way is built to the highest possible electronic and mechanical specification and is arranged to occupy the smallest possible space. It is then sealed up in epoxy resin, which permanently isolates the components in the

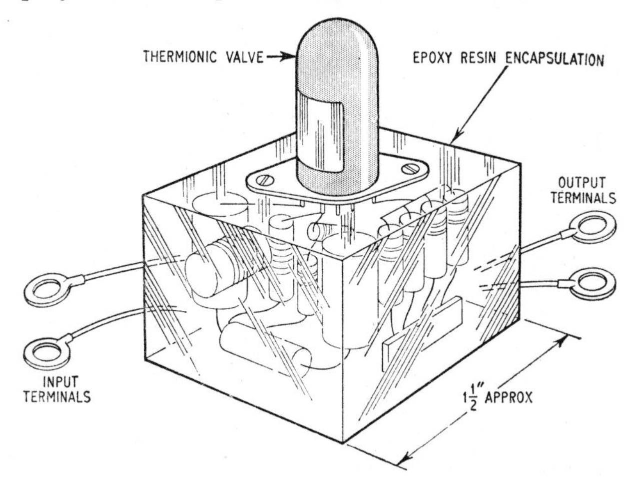

Fig. 1. A 'potted' amplifier circuit.

circuit from each other and holds them rigidly so as to reduce the effects of shock and vibration. If any component in the circuit fails the whole unit has, of course, to be replaced, but in the case of missiles this may not be a problem since the whole equipment may be expendable anyway. A typical potted circuit is shown in Fig. 1.

Printed Circuitry

Printed wiring is done by arranging the components to be connected in the circuit on one side of a suitable board, with their connection tags inserted through slots in the board to emerge on the other side, and preparing the tag side of the board in such a way that when it is plunged into a bath of molten solder and then withdrawn and cooled the tags are soldered to pre-prepared conducting paths which form the connections between the components. This is called 'printing' the circuit, and does away with the need for conventional wiring.

Extremes of temperature

The temperatures in space range from nearly one hundred degrees below zero to over a thousand degrees (Fahrenheit) above zero, nearly five times the boiling point of water. Most electronic equipment in space vehicles is transistorized (with the possible exception of the transmitter valve and some output stages), and transistors are notoriously temperature conscious. The best that the designer can do is to include temperature-compensating devices in the circuits and to ensure that the transistors work under conditions where the standing currents are kept to a minimum. This means that the transistors and valves work in a condition near to cut-off, i.e. in the absence of a signal of any kind they do not pass any current and therefore do not generate heat. On the ground this factor is not so important and it is common to find circuits which, for the sake of distortionless working, are designed to conduct current the whole time even when no signal is present. This creates unwanted heat but under normal conditions this can be blown away with a cooling fan. In space craft fan cooling is not possible because of the power needed to operate the fan and the fact that it takes up space which cannot be spared.

One of the problems of these high temperatures is their effect on the surface behaviour of metals—an effect which can, for example, cause relay contacts to stick permanently together.

Low pressure

Once out of the earth's atmosphere the pressure drops, there is a near vacuum, and this may cause sparks and arcing between leads and circuits. The electronic equipment is therefore sealed when installed to prevent this.

Cosmic and X-ray radiations

Short bursts of radiation produce spurious signals in semiconductor devices although only temporary physical damage is done to them. Long exposure to radiation may, however, permanently damage the components. Some degree of protection against radiation is possible if the circuits are potted in the manner already described.

The failure of Telstar late in 1962 was due to the effects of radiation on transistors. Bell Telephone scientists overcame this by making remote circuit adjustments initiated by special command signals.

Miniaturization of components

The miniaturization of electronic components for missiles has received a great deal of attention from designers over the last few years. The first step was the modular concept already used when a circuit is to be potted. The module can be regarded as a small, complete 'black box', something which usually has an input and an output, some electronic process being performed between these points.

Complex military equipment is generally made up of sub-assemblies—complete units, called modules, which can be removed and replaced whole with speed and accuracy. When such equipment is damaged the area of failure is quickly located and the malfunctioning module replaced whole, no attempt being made to analyse in the field what part of the module is at fault. This is known as the modular or unitized approach to design.

An early attempt at miniaturization of electronic circuits for missile work is shown in Fig. 2. The modules are mounted in

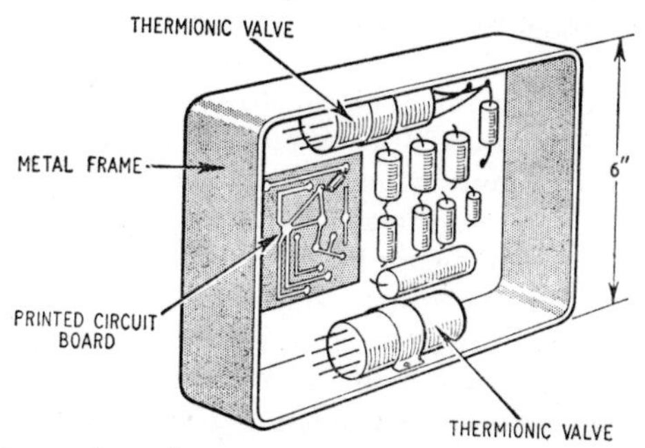

Fig. 2. An early type of miniature electronic module.

frames, the valves being attached to the frames to facilitate heat removal. The complete unit is plugged into a rack holding several such units which together form some apparatus. Servicing is simplified as frames can easily be withdrawn from the equipment rack for examination. A guidance system may use up to twenty of these modules. This technique is much used in computer construction.

Often in modular design a mixture of printed and conventional wiring is used, and thermionic valves and semi-conductor devices are mixed, depending upon the application or circuit function. With modular techniques it is possible to get up to 2,000 components

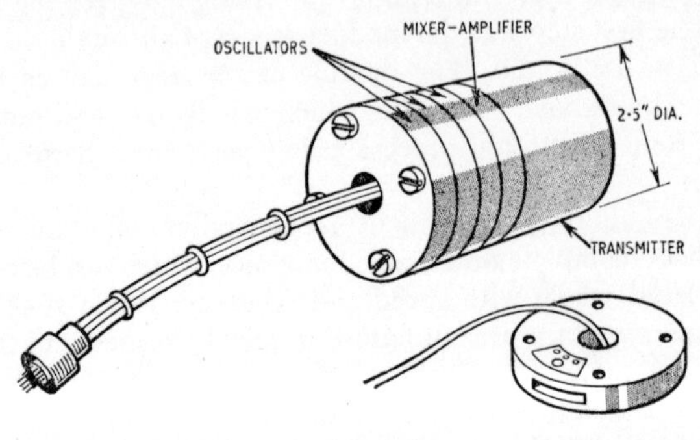

Fig. 3. Latest method of miniaturization: a telemetry system in modular form.

into a volume equal to an ordinary cutlery drawer—that is about one cubic foot.

It was subsequently found that many components as well as the wiring between them could be printed. Coils and capacitors can easily be made in this way and resistive material can be incorporated as necessary to form resistors. The latest techniques are to use carefully made semiconductor blocks: by controlling the junctions, etc. solid blocks can be 'grown' which fulfil the functions of quite complex conventional circuits.

Fig. 3 shows a typical example of miniaturization techniques for use in a five-channel missile telemetry system (telemetry = metering at a distance).

Power supplies in space craft

In the case of satellites which may have to remain in orbit for weeks, months or years the problem of finding power for the electronic equipment with which they are provided can be solved by using batteries. Since, however, about one thousand pounds of fuel is needed to put one pound of payload into outer space large numbers of batteries cannot be carried. Other means of obtaining power in satellites are therefore being experimented with by engineers. These include thermo-electric generators, nuclear sources and magneto-hydrodynamics.

Solar energy

One of the properties of a semiconductor pn junction is a photovoltaic effect. If either the n- or the p-type material is made thin enough to allow radiation to penetrate it the junction is irradiated. This has the effect of allowing holes and electrons to drift across

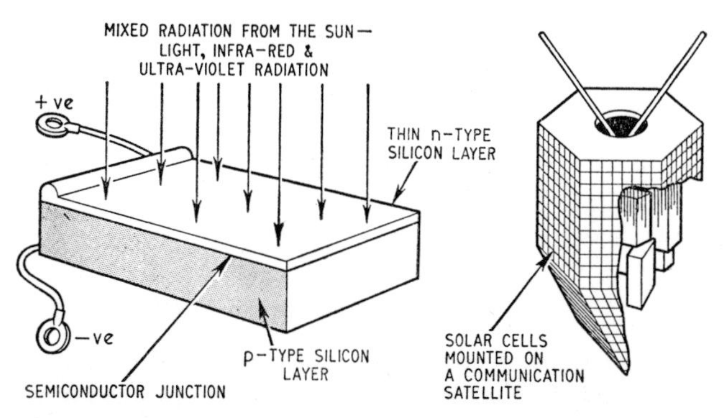

Fig. 4. The solar cell as a source of power.

the junction and thereby energy is released. A voltage is generated across the junction, the polarity being as shown in Fig. 4. This type of pn semiconductor arrangement is the basis of the solar cells used to provide energy in satellites.

The radiation from the sun is a mixture of visible infra-red and ultra-violet light. When this mixed radiation falls upon the cell the voltaic effect is produced. The open-circuit voltage of such a cell is of the order of about half a volt.

These cells are arranged on the surface of the satellite in the manner shown on Fig. 4 and are connected in series and parallel arrangements to supply the required voltages and currents.

Other sources of power

To reduce the payload in short life missiles various other types of power sources are under investigation. Among these is the thermo-electric generator, which has no moving parts and can be situated in the effluent of a missile or wherever there is sufficient heat to generate a voltage. The device is a development of the thermo-couple, which consists of two dissimilar metals joined in a closed loop. When one metal is heated whilst the other is cooled a current flows in the loop, a phenomenon known as 'Seebeck Effect'. With

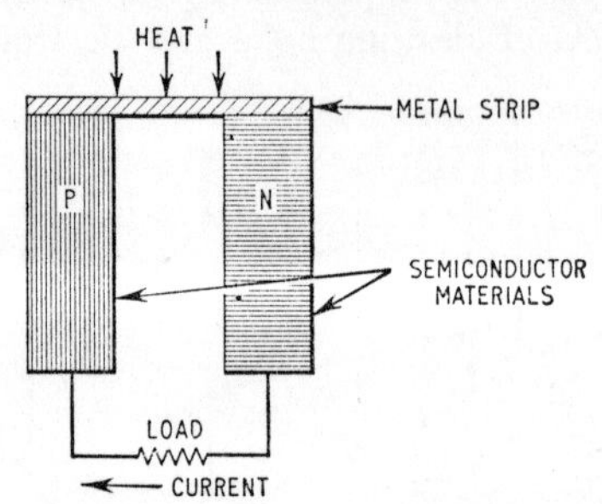

Fig. 5. Basic thermo-electric power unit.

the thermo-electric materials generally used such generators are not very effective—they are used mainly for temperature measure-ment. With the advent of semiconductor materials, however, much more efficient generators can be produced. Fig. 5 shows two semi-conductor elements coupled together, a p-type and an n-type. Neither of these materials need be as pure as the semiconducting materials used in the manufacture of transistor devices so that the production cost is below that of other semiconductor devices. The elements are made in the form of rods joined together by metal strips. A practical configuration is shown in Fig. 6.

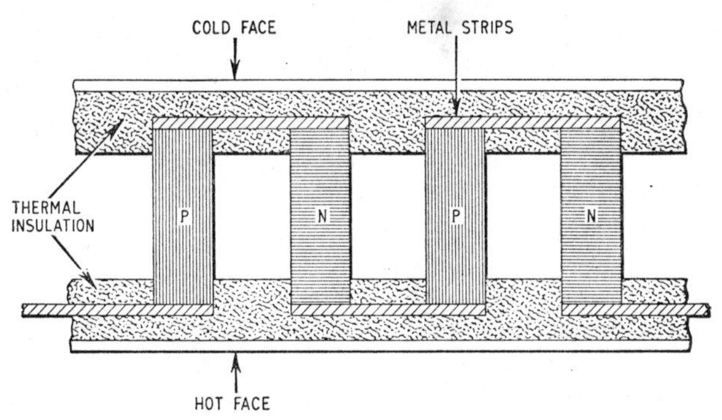

Fig. 6. A practical thermo-electric generator.

Another power source being investigated uses the principle of magneto-hydrodynamics. As shown in Fig. 7 a stream of hot gas containing ionized particles is passed through an intense magnetic field. As this gas stream is ionized it is effectively an electric current

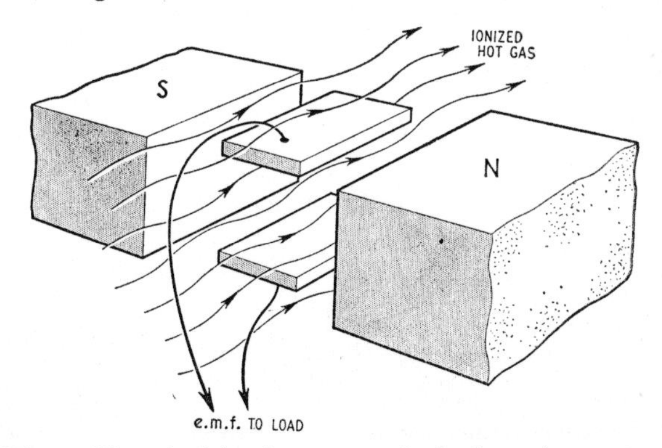

Fig. 7. The principle of a magneto-hydrodynamic generator.

flowing in a strong magnetic field and, in accordance with the laws of electricity, a voltage is produced across it as shown. If the electrodes are suitably arranged this voltage can be picked up and used as a power source. Although this principle offers a promising source of power for space craft better materials than are present available are needed before it becomes a practical proposition.

Telemetry in space craft

Information from satellites and space probes is sent back to earth via a radio link. There are many systems in use but all follow a similar basic pattern. See Fig. 8.

Inside the space craft a number of transducers sample the environmental conditions, e.g. temperature, cosmic radiation, pressure, infra-red radiation, presence of oxygen, ionizing layers, weak magnetic fields and so on. If the missile is being used for military purposes it may make other measurements and take infra-red photographs, etc.

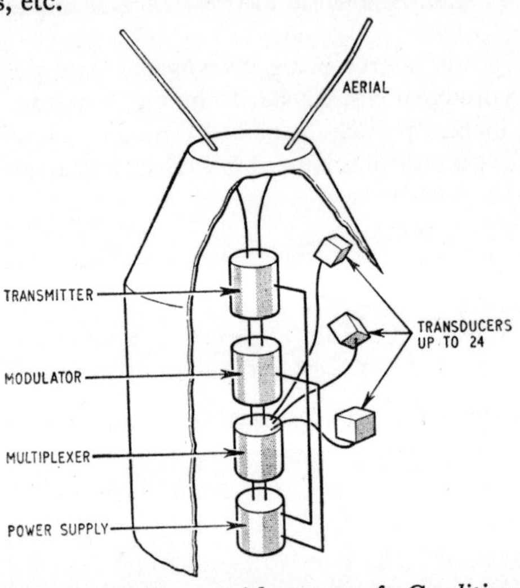

Fig. 8. Telemetry system for use with space craft. Conditions in the space craft are measured by the transducers and sent back to earth via the transmitter.

The measurements may be made with a variety of transducers from variable inductors to piezo-electric accelerometers. Semiconductor cells fitted in special chambers on the surface of the craft measure radiation and particle bombardment and a magnetometer of sufficient sensitivity can be used to measure weak magnetic fields.

Each transducer produces an electrical signal the frequency and amplitude characteristics of which are analogues of data being measured. These electrical signals are then sent back to the ground station, via a radio link, for assessment.

Radio link

The signals cannot be sent back direct: they are too weak. The following technique is therefore used. They are first amplified and then used to modulate the output of a powerful high frequency oscillator. This oscillator, working at a very high frequency, produces what is called a *carrier wave*, i.e. a wave which when modulated 'carries' the information from the radio transmitter to a radio receiver. The modulated carrier wave is fed to a suitable aerial and transmitted back to the ground station. There it is received and the original signals are detected, that is they are extracted from the carrier wave.

A further problem is the transmission of the simultaneous outputs of a number of transducers. Clearly if we have, say, twenty transducers we would be in serious difficulty if we had to have twenty transmitters to send their signals back to earth. This problem is overcome by *sampling* the output of each transducer in turn: a small amount of each second of transmission time is devoted to each transducer in turn.

Inside the space craft a multiplexer, that is a motor-driven switch, is connected to each transducer in turn, linking it to the transmitter. Typical rotating speeds for the switch are 100 times per second to as low as once per second. Each time the switch is connected to a particular transducer it feeds the output signal of the transducer to the transmitter modulator together with a synchronizing signal provided by the multiplexer to tell those at the receiving station which transducer is connected at that moment.

The modulator is a special circuit which varies the carrier wave generated by the transmitter's h.f. oscillator in such a way that the information to be sent is imposed on the carrier wave. This modulation may be done by altering the frequency of the oscillator or the amplitude of the carrier wave. An alternative arrangement is to send the information in the same way that the morse code is sent—by switching the carrier on and off in accordance with the information to be sent. The preferred methods are to modulate the oscillator frequency (*frequency modulation*) or to use the morse code technique. These methods are less liable than modulation of carrier amplitude (*amplitude modulation*) to interference problems.

Receiving the signals from space craft

On the ground a sensitive receiver picks up the signals and by using a de-multiplexer with associated equipment translates the signals to their original quantities for assessment. The synchronizing pulses produced by the multiplexer are used to assist the de-multiplexer in sorting out the various signals, which after detection (demodulation) may be immediately displayed or stored on magnetic tape for future investigation.

Further instrumentation

The space craft, missile or satellite contains a great deal more instrumentation then we have so far considered. Some is electronic and some electro-mechanical. Much is used for guidance and the tracking of the vehicle. Hybrid combinations of accelerometers and gyroscopes both control and measure the speed and direction of the craft and correct it by means of servomechanism systems. Transponders, transmitters that operate only on the receipt of a signal from the ground, transmit radio frequency signals to tracking stations where, by comparison of the phase of these signals, calculations, using computers, can be made so that the movement of the craft can be followed. Much of this instrumentation is the province of the electromechanical engineer but his problems are the same as ours—to reduce the payload by miniaturizing as many of the instruments as is possible consistent with efficient performance.

Space age timing mechanism

An example of the need for accuracy and compactness in space instrumentation is the timing mechanism used in some vehicles for the time switching of electronic apparatus. This requires a highly accurate mechanism that must for obvious reasons be small in size. The normal clock-mechanism with its balance wheel and hair-spring can be made small enough to act as a tiny time switch but it is not very accurate—the best may vary by several seconds a day. In addition they are liable to error from shock, which is especially high when a missile is launched.

An American firm has produced a transistorized clock mechanism which is extremely accurate and is guaranteed to an accuracy of two seconds per day. It was originally invented by a Swiss, Max Hetzel, in 1954, and has only twelve moving parts as opposed to the 26 needed for the normal mechanical clock mechanism. As it has no balance wheel and no hairspring it is not so easily disturbed by mechanical shock. The movement, which is only about an inch in diameter, combines the skills of the electronic engineer and the watchmaker.

The mechanism consists of a tiny, accurate tuning fork, oscillating at 360 times per second, to which are attached small magnets, coupled to the coils of a transistor pulse generator. The vibrations of the tuning fork are compared with those produced by the pulse generator and corrected if they change in phase or frequency. The transistor circuit is powered by a tiny mercury cell about the size of an aspirin.

9
TELEVISION

Television is the transmission and reception of moving pictures over a distance. Like the cinema, television makes use of the persistence of vision: if the information to form the pictures can be assembled sufficiently rapidly on the screen being viewed, the viewer will see a complete picture.

When light rays enter the eye and strike the retina at the back of the eye the impression they make does not cease immediately but remains for some time afterwards. This persistence of vision is long enough for separate pictures, following one another in time, to give the impression of continuous movement. Each picture follows its predecessor rapidly enough for the eye still to be occupied with the previous one, which it replaces. The impression of movement is thus obtained from a series of still pictures. In the television studio the scene to be transmitted is first 'photographed' at intervals by the television camera.

Scanning

Before transmission each of these still pictures must be reduced to a series of electrical signals. This is done by scanning the picture. The scene, by being scanned from left to right in a series of lines, is broken up into a series of light impressions of varying intensity. These are reassembled at the receiver to form the picture seen by the viewer.

The camera

There are a number of different types of camera tubes in use. These range from the relatively simple vidicon tubes used for outdoor work to the latest and undoubtedly the best indoor camera tube, the $4\frac{1}{2}$ in. image orthicon. Fig. 1 shows in simplified form the image orthicon at present used in television studios. The scene to be televised is focused on to the photocathode by a normal optical system. Under the influence of the light from the scene the photo-cathode emits electrons in quantities which depend on the amount of light falling upon it—those parts of the scene which are bright cause more electrons to be emitted than those which are dark. These

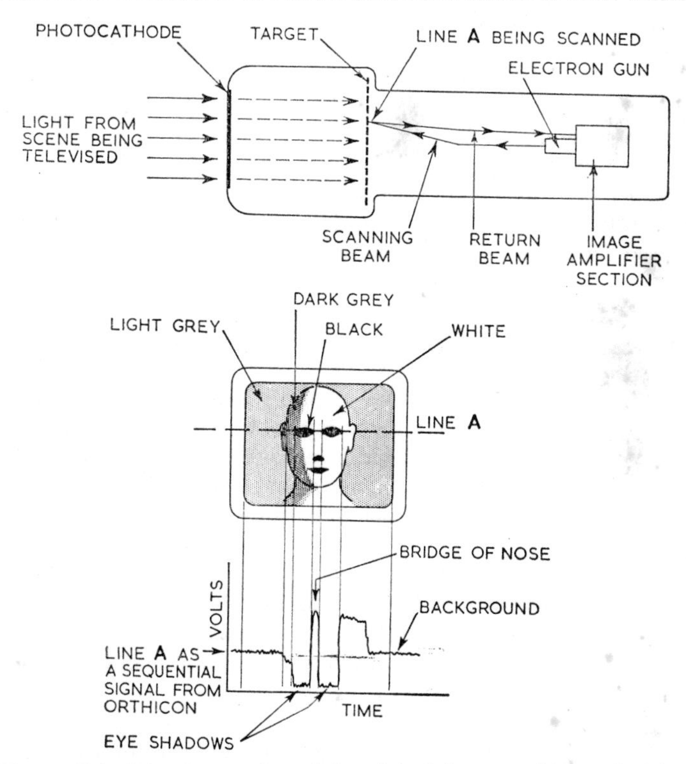

Fig. 1. Principle of operation of the 4½-inch image orthicon television camera tube and the way in which a scene is scanned to provide a sequential electrical signal.

electrons are attracted to the target where in this way an electronic image of the scene is produced, the image changing continuously as the original scene changes. At the other end of the tube an electron gun emits an electron beam which scans the target, rather as the eye scans the page of a book. In doing so it samples each part of the electronic image. Some of the electrons of the beam return, forming a return beam, others remaining on the target (to neutralize the electrons given off by the photocathode). The return beam is thus varied in strength, depending on the electrical value of the electronic image at each point. For instance a strong light on the photocathode produces a large electric charge on the equivalent part of the target and when the beam samples this point a large change is produced in the return beam. The return beam passes through an image amplifier which increases the strength of the signal by up to 2,000 times. The camera tube thus provides a sequential series of electrical signals of value equal to the brightness of the various parts of the picture.

Reassembling the picture

After further amplification these signals are transmitted. If picked up over distances of many miles they will be received in a weakened condition. They thus require amplification before they can be applied to the receiver cathode-ray tube. The signals vary the strength of the cathode-ray beam in the tube in the domestic receiver in an exactly similar manner as in the camera tube and as the beam scans the face of the tube the picture is reassembled.

Breaking up the picture

The speed at which the picture is scanned is such that a complete picture is produced every 1/25th of a second. This is sufficient to give the viewer the impression of watching a continuous event.

Interlacing

To enable the electron image on the target of the image orthicon to be scanned the scanning beam must be deflected—both horizontally and vertically. Coils are wound round the tube and by applying suitable waveforms to them varying fields are produced in the tube to deflect the electron beam. These waveforms are provided by the timebase circuits, the frame (or field) timebase deflecting the beam vertically and the line timebase deflecting it horizontally. In the television receiver the cathode-ray tube is deflected in exactly the same way to trace out pictures. Depending upon the system 405, 625, 819, etc., lines are scanned per picture ('frame') and there are 25 frames per minute in most systems. It has been found that if the frames are split so that half the lines are scanned in 1/50th second and then the beam returns to scan the alternate lines during the next 1/50th second, then although a complete frame is still scanned only once every 1/25th of a second the eye will be given the impression of seeing a frame every 1/50th of a second and flicker is reduced to negligible proportions. This technique is called interlacing.

The synchronizing pulse generator

In the television receiver and in the camera electrical signals are needed so that the deflections of the electron beams in the camera and the receiver cathode-ray tube are synchronized. These signals are provided by the synchronizing pulse generator, which is shown connected in the system in Fig. 2.

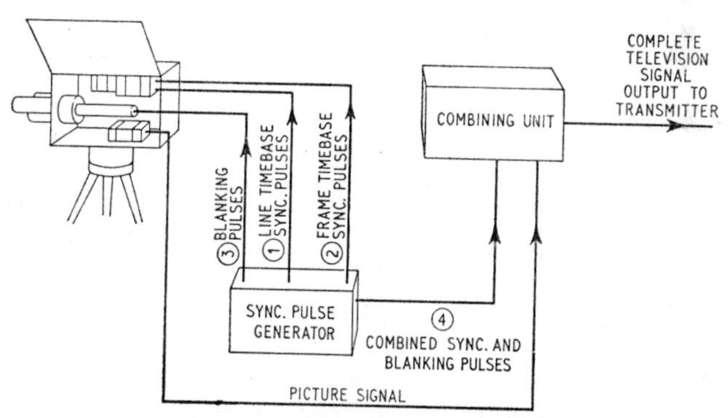

Fig. 2. Sync. pulse generator used at the studio.

The sync. generator, as it is called, consists of a 'master-oscillator' and a number of counting-down, gating and mixing circuits. It has four outputs: (1) A series of pulses which are fed to the camera tube line timebase to control the horizontal scanning of the image orthicon target in a regular manner. These are called the line sync. pulses. (2) A series of pulses which act in the same way to control the vertical scanning of the camera tube target. These are called the field or frame sync. pulses. Both these pulses are a little ahead in time of the other pulses from the generator—the camera tube has to start operating before the rest of the equipment. (3) A series of blanking pulses. These are fed to one of the camera tube electrodes and virtually switch the tube off during the time when the electron beam is flying back to the start of a line or frame in order to suppress the flyback traces. (4) Combined sync. and blanking

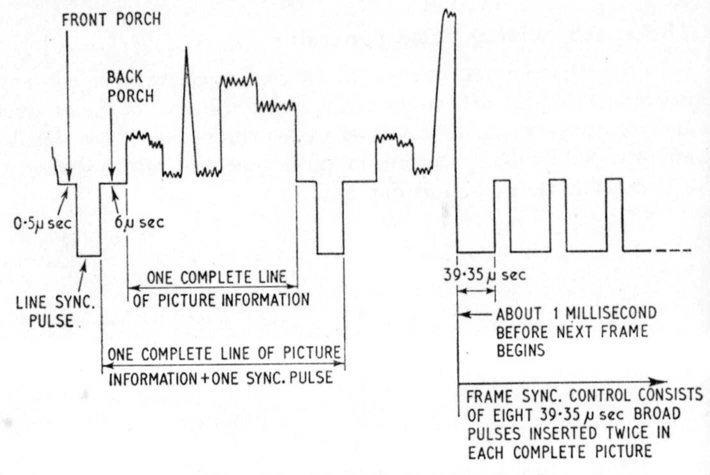

Fig. 3. The composite television signal.

pulses. This output is fed to a mixer unit where it is combined with the picture information to make up the composite television waveform shown in Fig. 3. The composite series of signals is then fed to the transmitter with the scanning information included in it to enable the domestic receiver to scan in synchronism with the camera tube in the studio.

The composite television signal, called the video signal, may be connected to Post Office lines, or, in some cases, to a radio link, before being fed to the transmitting station, which may be many miles away from the studio. The sound signal is separate at this stage, being obtained by using the same techniques as for ordinary sound broadcasting—microphone, amplifiers and so on—and fed to the transmitter over a separate line.

The transmitter

Transmitting stations are usually located as high above sea level as possible and also central to the area they are to serve. This often means that they are situated in rather remote places. In the station, apart from the control gear and monitoring equipment, there are transmitters for both the vision and the sound signals. In practice there are two transmitters for each purpose in case of breakdown. The main vision transmitter may have a power output of the order of 5 kilowatts and the sound transmitter an output somewhat below this figure, perhaps 1·5 kilowatts. The video and sound signals from the studios are sent to the transmitting station and then fed to the modulating stages of the transmitters where they vary the amplitude of the transmitter carrier wave output (see description of radio transmission on page 145). Both the sound and vision transmitter outputs are then fed to a combining unit which matches them both to the transmitting aerial.

The aerial system consists of an array of dipole elements connected together to give either a directional or an all-round, omnidirectional transmission depending upon the location of the transmitter and the area to be served.

The receiving aerial

The combined signal may be received on an aerial array of the type shown in Fig. 4. The value of the signal after travelling about 30 miles may only be of the order of millivolts. The aerial shown in Fig. 4 has three elements: the active one, called a dipole, to which the down-lead to the receiver is connected, a director and a reflector. The director concentrates the signal on to the dipole and in order to tune it is made slightly shorter than the dipole. It must also be placed a specified distance in front of it. The reflector is longer than the dipole and is situated a specified distance behind the dipole. The polar response of such an aerial display is shown in Fig. 4. It can be seen that nothing is received from directly behind, that is where the reflector is situated, the direction of maximum gain being where the director element is situated. Normally the aerial is pointed towards the transmitter. However, in cases where there is considerable interference, i.e. 'noise', it is often better to point the zero position of the aerial towards this interference and pick-up slightly less signal from the transmitter. In this way a better signal-to-noise ratio is obtained. The more directors included in an aerial

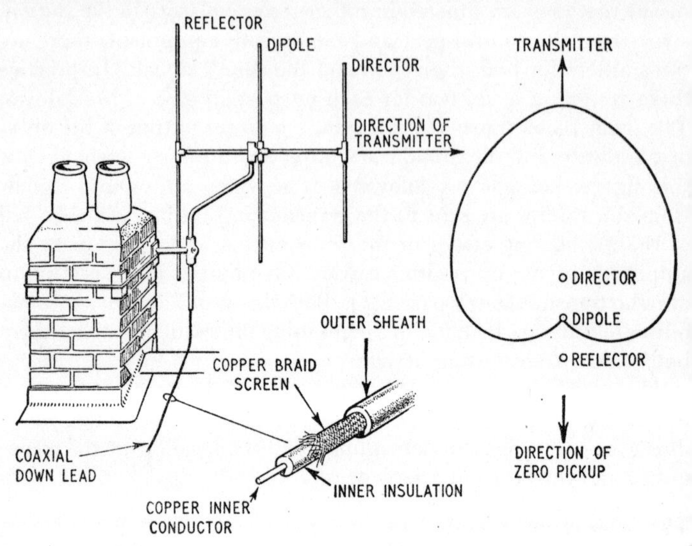

Fig. 4. Aerial for television reception. The polar response of the aerial is shown on the right.

array the sharper and therefore more discriminating is the resultant polar diagram. In the case of Band III (I.T.A.) transmissions, where the frequency is much higher than the B.B.C. transmission on Band I, a number of directors can easily be fitted to the aerial system as the physical size of the aerial is much smaller (the higher the frequency the smaller the dipole—the length of the dipole should be approximately half that of the wavelength to be received). This accounts for the different appearance of B.B.C. and I.T.A. aerials although they work on the same principle.

The receiver

The signal from the aerial is fed into the receiver by means of a coaxial cable as shown in Fig. 4. This cable is specially constructed to match the aerial system to the receiver input and also to screen the signal from stray voltages which may otherwise be picked up by the down-lead. The metal sheath around the coaxial cable provides this screening.

Fig. 5 shows the basic television receiver in block form. The first stage is an amplifier capable of handling both the vision and sound signals. This amplifier is incorporated in the tuner, which

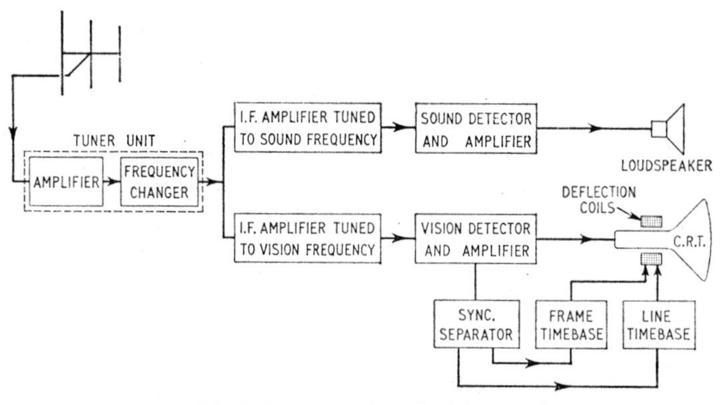

Fig. 5. Block diagram of a television receiver.

enables the correct channel selection to be effected. The initial r.f. amplifier stage is followed by a frequency changer stage which changes the carrier frequency of the selected and now amplified channel to a much lower, fixed frequency called the *intermediate frequency*. This is always the same whichever channel is chosen. Reducing the frequency makes the design of the following i.f. amplifiers needed to further amplify the signal much simpler. The output of the i.f. amplifiers is fed to the sound and the video detectors. Tuned circuits are incorporated in the i.f. stages to separate the sound and vision signals. The detectors remove the carrier wave, leaving just the sound and vision signals which are applied to suitable output stages for final amplification. The sound output stage drives a loudspeaker. The vision output stage feeds the picture tube and also the sync. circuit where the sync. pulses are removed from the composite waveform. The sync. pulses are then used to control the timebase circuits which, in turn, control the scanning of the cathode-ray tube. The timebase circuits control the *movement* of the electron beam in the cathode-ray tube, the output of the video output stage *modulating* the beam so as to control the *brightness* of the spot it produces.

10

ELECTRONICS IN INDUSTRY

The use of electronics in industry has greatly increased in recent years. Electronic techniques can be used to measure thickness, temperature and strain, for weighing and counting, to control drying processes, detect radiation, measure flow, control electric motors and so on. Computers can take the place of supervisory staff to control complex operations and may be used to work out the cost and possible profitability of proposed industrial schemes.

This chapter gives a few examples of the industrial use of electronics to show the useful role that electronics plays in modern industry.

Electronic counting

There are a number of ways in which electronic techniques can be used in factories where large quantities of small items, for example small metal parts, have to be counted.

One method uses an ordinary microphone as shown in Fig. 1. The small metal parts drop singly on to a platform under which a

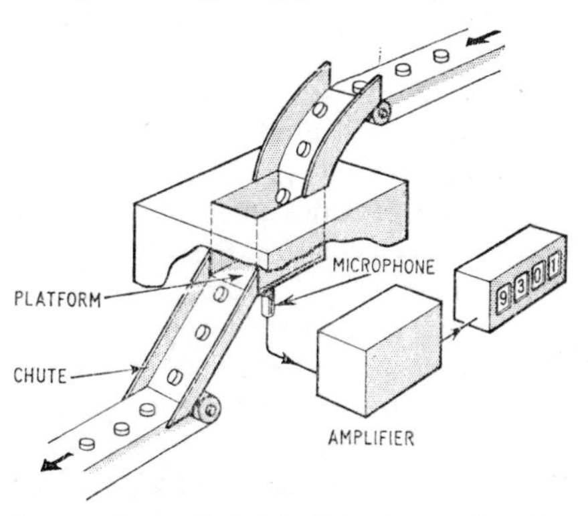

Fig. 1. Counting small metal articles by using a microphone and counter.

microphone is situated. The microphone records the sound as each item hits the platform prior to slipping away down the chute to a collector or to a moving belt to be carried away for packaging. The microphone output is connected to an amplifier the output of which is fed to an electromechanical counter which records the number of sounds picked up by the microphone.

A more sophisticated method which may be used where larger objects are to be counted or where the objects are fragile or cannot be handled uses a photocell as shown in Fig. 2. The photocell is placed opposite a light and as objects pass they interrupt the light beam. When the light in the photocell is thus interrupted the

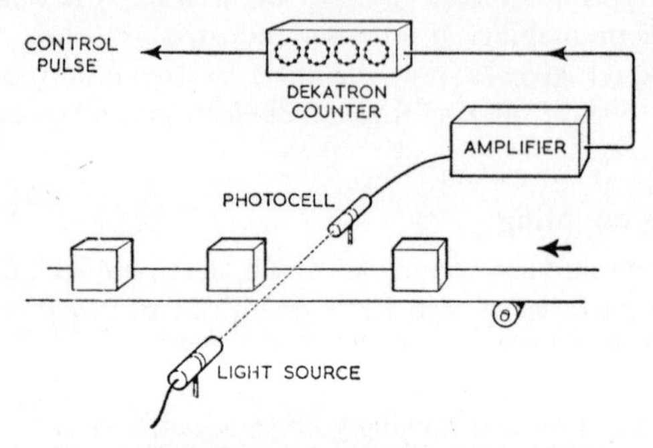

Fig. 2. Using a photocell to count packages.

photocell is inactivated. Each time this occurs the photocell applies a signal to the amplifier which produces a pulse that is fed to an electromechanical or electronic (e.g. a dekatron) counter. It can be arranged that the counter provides an output signal after a certain number of objects have been counted, the output signal being used to initiate some further action such as stopping the conveyor belt.

As an extension of this it is possible to use the counting rate to control the time of some process such as drying or curing. For example if electronic potted circuits are counted after passing through a curing oven to harden the potting resin the rate at which they are counted is a function of the time they spend in the curing oven. If this information is correlated with the oven temperature it is possible to ensure, by means of suitable control equipment, that each potted circuit spends the correct time in the oven.

Measuring liquid levels

There are a number of ways of measuring the levels of liquids in tanks and other containers. One common method uses a capacitance probe the capacitance of which changes as the level of the liquid rises and falls. This change unbalances a bridge circuit (see Chapter 5), producing a signal which is used to give an indication of the level on a suitable meter connected across the bridge.

Another simple device uses three probes as shown in Fig. 3. One

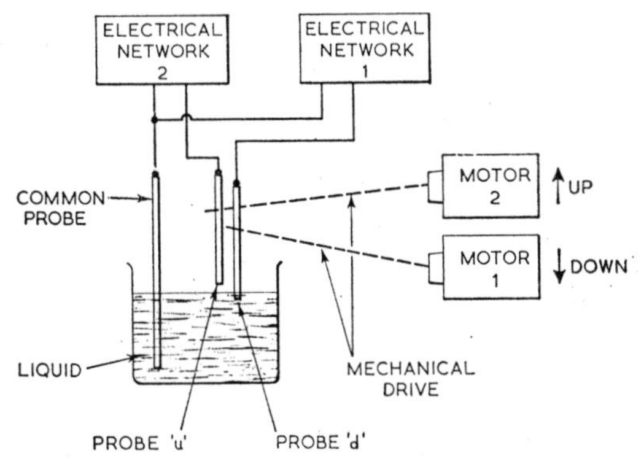

Fig. 3. An automatic liquid level indicator.

common probe is in a fixed position permanently in the liquid. The other two are driven up or down by two motors. Suppose that the longer one, *d*, is driven downwards by the down motor (1). When it touches the surface of the liquid, a d.c. current will pass between it and the common probe via the liquid. The current is used to activate a relay which switches off the motor driving the probes down and the probes stop. This state of affairs will continue until the liquid level changes. If it falls probe *d* is again exposed, the relay circuit is broken and motor (1) drives the probes down until *d* touches the liquid again. If, however, the liquid level rises then the third probe *u* will come into contact with the liquid and a second circuit is completed (between this and the common probe). The current flowing activates another relay which starts motor (2). This drives both probes up. When probe *u* is out of the liquid this second circuit is broken and the second motor stops. Thus as the liquid moves up and down motors 1 and 2 are energized to correct the position of the probes until the resting condition is reached, with probe *d* just dipping into the liquid and probe *u* just out of it. Signals derived from the motors can be converted and used as an indication of the level of the liquid.

Measuring flow

Measuring the flow of liquids in pipes with no restriction to the flow of the liquid can be achieved by means of the arrangement shown in Fig. 4. A coil is wound around the pipe carrying the fluid

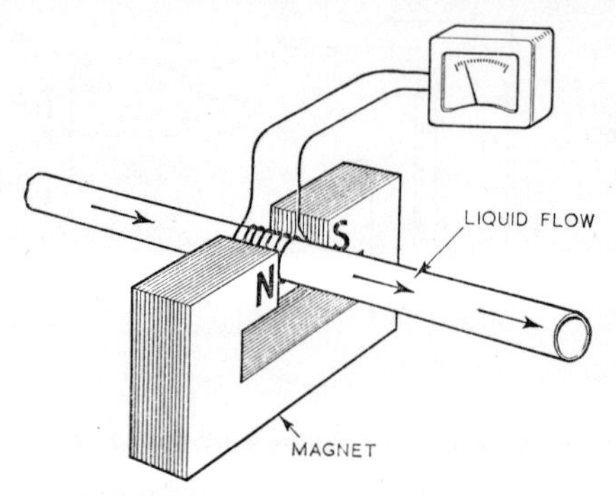

Fig. 4. Measuring the flow of liquid in a pipe.

and a large magnet is placed across it. The liquid, which must of course be electrically conducting, flows through the magnetic field so that a current is induced in it. This is coupled to the pick-up coil which provides an indication on a meter. The deflection on this meter is directly proportional to the amount of liquid flowing and thus the meter reading is a measure of the rate of flow.

Another type of flowmeter, which can be used when the liquid is non-conducting, is shown in Fig. 5. This is a more expensive equipment and uses ultrasonics in measuring the flow. Two barium titanate crystals are placed in the pipe diagonally. Ultrasonic energy is fed to one crystal at a time while the other is used to receive the ultrasonic energy. In this way the time taken for a pulse of ultrasonic energy to travel between the crystals can be measured. If the pulse is fired against the fluid flow it will take longer to reach the

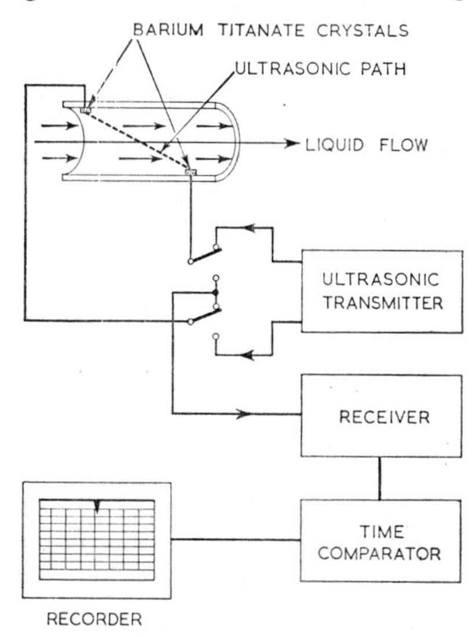

Fig. 5. An ultrasonic flowmeter.

receiver crystal than if it is fired downstream. The difference in these times is directly proportional to the rate of flow. The pulses are fired alternatively from each crystal, and picked up by the other, the times being compared in an electronic comparator circuit. The final output, which is proportional to the flow rate, may be recorded.

Moisture content

A relatively simple technique can be used to test materials like cloth and paper for moisture content. The material to be tested is passed over rollers between two metal plates which act as the plates of a capacitor. The material is thus the equivalent of the dielectric of a capacitor.

If the cloth contains moisture the value of the capacitance will be reduced, and vice versa. If this capacitor is made one arm of a measuring bridge then the amount of moisture present in the cloth will be converted into an electrical signal which can be used to give

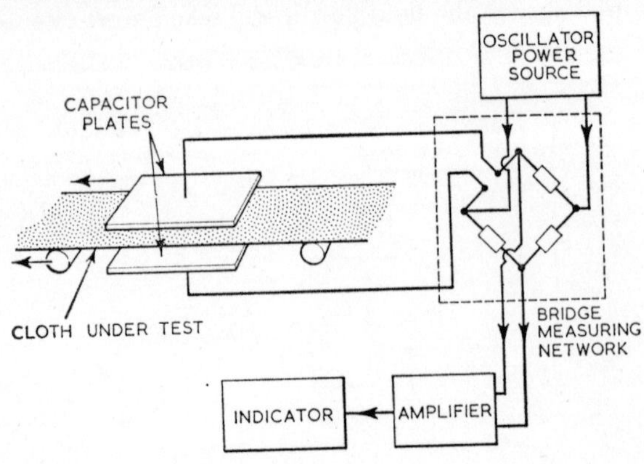

Fig. 6. Method of electronically measuring the moisture content of cloth.

an indication or warning or take some action such as increasing the temperature of driers. Fig. 6 shows a possible arrangement for checking moisture content of cloth being passed over rollers.

In agriculture a similar method is used to dry out moist grain. A large specially-shaped capacitor is situated in the mass of grain

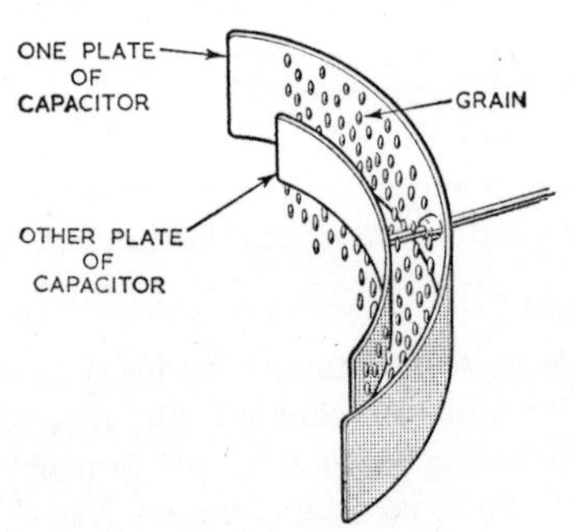

Fig. 7. Grain drier sensor.

which is passing through the drier. The change in capacitance forms a signal which is applied to a servomechanism. This alters the rate at which the grain travels through the drier to ensure that it is evenly dried. The capacitor used in this application is shown in Fig. 7.

Ultrasonics in industry

Sound waves are longitudinal disturbances in a medium such as air or water. Electromagnetic waves differ in being transverse in character. The energy changes of transverse waves occur in a plane vertical to the direction of movement, rather like ripples on a pool of water when a stone is thrown in, whereas sound waves are alternate rarefactions and compressions of the medium in which they travel, which of course must be elastic in order to transmit them.

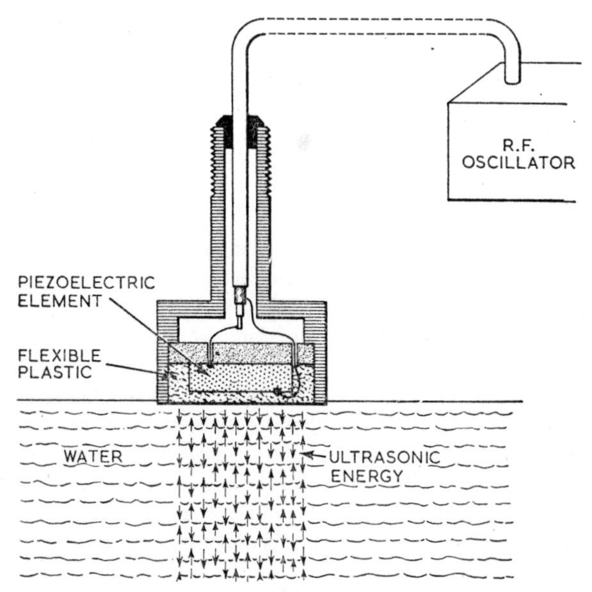

Fig. 8. Transducer producing an ultrasonic disturbance in water.

Ultrasonic waves are the same as sound: they are not, as some people imagine, like radio waves. The confusion arises because they are high in frequency—they may even be in the megacycles per second region. Even so they are totally unlike the radio frequency waves that are used to produce them. The term ultrasonic means 'above sound' and that is precisely what they are. Many people cannot hear any sound with a frequency higher than about 8,000 c/s so that to them anything above this value would be ultrasonic. Generally speaking, however, it is usual to regard ultrasonic waves as being those above about 16,000 c/s.

To produce an ultrasonic disturbance in water or a similar medium some form of transducer is needed. This may be made of barium titanate—a material which has piezoelectric properties, that is, it oscillates mechanically when electrical oscillations at the right frequency are applied across it, and vice versa.

A typical ultrasonic transducer is shown in Fig. 8. The leads are coupled to a source of radio frequency energy of appropriate power. A transducer such as that shown needs about as much power to operate it as a large electric light bulb. Ultrasonic oscillations are transmitted into the medium, in this case water, and if the transducer is correctly shaped the energy will be transmitted as a beam.

Detecting flaws with ultrasonics

In the iron founding industry ultrasonics have assisted in making quality testing a much more exact task. In the past, large iron castings weighing many hundredweights and costing many hundreds of pounds were sold to customers for machining, a process that might take several weeks, and it was quite possible that at the end of this time, whilst a final cut with a milling tool was being made, a large hole would be revealed in the block due to faulty casting. This meant not only waste of the casting but serious loss of time. Today, however, although flaws still occur in large castings—in fact it is impossible to guarantee that they will not sometimes occur—they can be detected before the casting is sold so that loss of time and money is avoided.

This quality checking is done by an ultrasonic flaw detector. The transducer head is coupled to the casting as shown in Fig. 9,

sing some water-bound cellulose paste, and the oscillator produc-
ng the energy to power the transducer is switched on. This oscilla-
or is pulsed (as in radar) so that it produces short sharp output
ulses at the rate of perhaps several hundreds per second, each
ulse lasting for perhaps a few milliseconds. The pulses are fed to
he transducer and transmitted through the casting, and echoes
vill return along the path of the beam. These echoes are picked up
y the transducer, reconverted to electric signals, and can be dis-
layed on an oscilloscope. Echoes occur whenever the ultrasonic

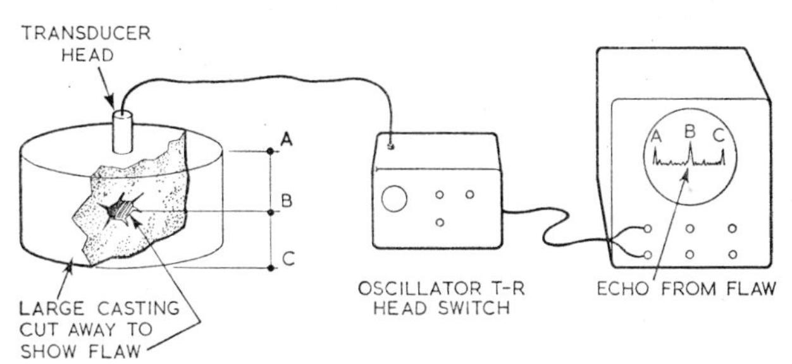

Fig. 9. Ultrasonics used to check for flaws in metal castings.

eam encounters an alteration, e.g. a flaw, in the medium through
vhich it is being transmitted. In this way the presence of flaws will
how up as echo signals on the oscilloscope.

The same transducer head acts as both the transmitter and
eceiver of the ultrasonic energy: after each transmitted pulse the
ead is electronically switched over to 'receive' to await the return
ulses from the casting. Fig. 9 shows a typical set of echoes from a
asting, echo A being caused by the interface between the trans-
ucer and the casting, B by a flaw in the casting and C by the
nterface between the casting and the surface on which it stands.

lectronic measurements

By using suitable transducers to convert various parameters, e.g.
emperature, pressure, etc., into electrical signals, a large number of
uantities can be electronically measured. The transducer or sensor
s usually included in one arm of a measuring bridge, the bridge

being the most sensitive method of detecting small electrical variations. There are numerous bridge arrangements for measuring different electrical quantities, e.g. resistance, capacitance, etc., but all are basically the same and an understanding of the action of the simplest of them will enable the reader to appreciate the working of the others.

The Wheatstone bridge

The Wheatstone bridge for resistance measurement is shown in Fig. 10. Three of its arms contain resistors of known value whilst

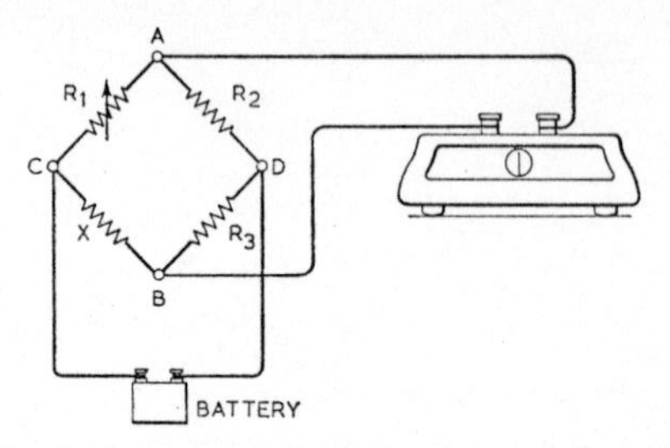

Fig. 10. The Wheatstone bridge for the accurate measurement of resistance.

in the fourth is a resistor whose value is unknown. A small battery is connected across points C and D, and a sensitive meter is connected across points A and B to measure the current flowing in the circuit. On first connecting the apparatus a current flows through the meter since the bridge is unbalanced. Resistor $R1$ is then varied until the galvanometer reads zero. When this happens the bridge is said to be balanced and the ratio of $R1$ to X, the unknown resistance, is the same as the ratio of $R2$ to $R3$. A simple algebraic calculation will then give the value of the unknown resistor. In practice the apparatus is usually so calibrated that the value of X can be read directly from a dial attached to the control knob of $R1$.

The bridge method is a very sensitive way of measuring resistance values and can easily be adapted for industrial purposes by using as the unknown resistor a transducer sensitive to pressure, temperature, flow or any other variable quantity. By using a bridge and sensitive meter very small changes in the value of these quantities can be detected.

The galvanometer

This is a much more sensitive instrument than the moving-coil meter and is used as the null detector in bridge arrangements. The ordinary moving-coil meter will read zero even when there is a small current flowing because the movement is not sufficiently sensitive to move under the influence of tiny currents. The galvanometer, being much more sensitive, will move under the influence of much smaller currents.

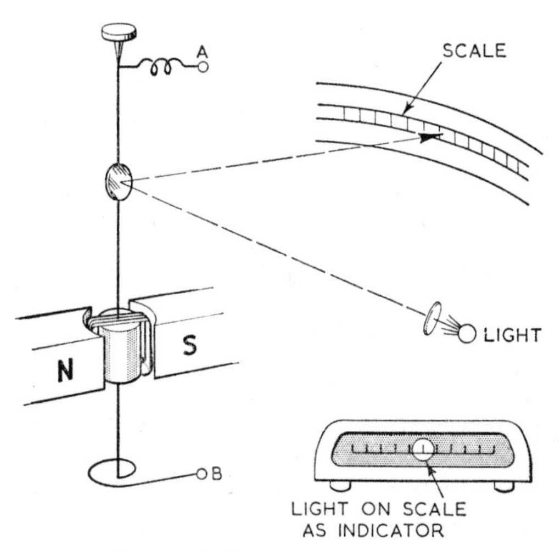

Fig. 11. Galvanometer movement.

The high sensitivity of the galvanometer is due to the way in which its movement is mounted, a system called single suspension. Unlike the moving-coil meter the movement has no pivots and therefore has a very low torque which enables it to respond to very small currents. In Fig. 11 a moving-coil galvanometer with single suspension is shown. N and S are the poles of a magnet between which the moving coil is suspended. The electrical circuit is from

A to B through the coil. When a current is passed through the circuit the movement twists due to torque. A mirror is attached to the suspension and reflects a light beam focused on to it along a scale at the front of the galvanometer. Some amplification of the meter movement is obtained by this optical method of indication and this of course further increases the sensitivity of the galvanometer.

Measuring temperature with a thermistor

The bridge method with a galvanometer as a null detector can be applied to most industrial measurements. An example is temperature measurement using a thermistor (see Chapter 3) as the unknown resistance. As temperature changes so the resistance of the thermistor changes and the bridge becomes unbalanced. The bridge can then be rebalanced by changing the value of $R1$ (Fig. 10). As the resistance/temperature characteristics of the thermistor are known it is possible to calibrate the dial in temperature directly in degrees centigrade.

In practice it may not be convenient to rebalance the bridge manually each time it is unbalanced by a temperature fluctuation. A d.c. amplifier may be used to detect the bridge unbalance and the output of the amplifier fed to a servomechanism which rebalances the bridge automatically. The movement of the servomechanism is an indication of the degree of unbalance and hence of the temperature variation. This more complex system enables remote indication of temperature changes to be given. If necessary the arrangement can be extended to give warning if the temperature changes beyond some specified limit, or to supply a signal which can be used to adjust some operation or process that will modify the temperature.

This method may also be applied to the measurement of pressure, displacement and so on.

Automation in industry

Electronic techniques have made possible great advances in automation in the last few years and there is today hardly a plant in the country without some automated process. In the chemical industry for example some large refining plants are run for months with only a handful of staff, most of the processes being automated.

Fig. 12 shows in outline a basic automatic control system.

Suppose that we wish a conveyor belt to travel at a certain speed. This is represented in Fig. 12 as the 'operation under control'. The conveyor belt is driven by an electric motor. To this is linked a device providing an electrical signal which is proportional to the motor's speed. This signal is compared by the comparator with a *reference signal* corresponding to the speed at which the motor should run to drive the conveyor belt at the desired speed. The comparator provides an *error signal* corresponding to the difference

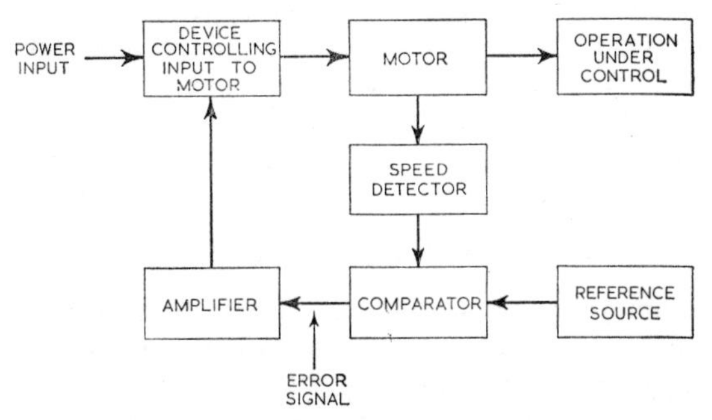

Fig. 12. Block diagram showing the basic elements of a closed-loop automatic control system.

between the actual speed of the motor and its correct speed, and this error signal, after amplification, is used to control the power input to the motor so that the motor's speed is automatically compensated. This is called a 'closed-loop' feedback control arrangement: information on the actual operation under control is fed back and used to adjust it automatically. There is, of course, always some slight error in the system because deviation cannot be corrected until after it has taken place, but this error can by careful design be kept very small.

Thyratrons and silicon controlled rectifiers are electronic devices that are frequently used for the control of electric motors. Where a d.c. motor is powered by an a.c. supply after rectification, they can be conveniently used to provide the necessary rectification. In such cases the error signal can be used to control the gate of a silicon controlled rectifier or the grid of a thyratron used as a rectifier, thereby controlling the power supplied by the rectifier to the motor.

Feedback control arrangements can be devised to control temperature, humidity, flow, pressure, weight—in fact almost any factor which can be electronically measured by means of a suitable transducer. Also, if the reference source is 'programmed', i.e. takes the form of a series of different signals, then the operation under control can be made to perform a series of operations automatically. A programme can be provided by a computer or a pre-recorded tape, and used in the control, for example, of machine tools.

Electronic systems of automation relieve man of the tedious tasks of measuring, assessing and correcting equipment, and can operate at much faster speeds.

COMPUTERS

The complexity of calculations associated with modern industry, commerce and research has in many cases increased to a point where solution by normal manual techniques is extremely tedious, difficult or impractical. Electronic computers are being increasingly used to solve such problems. There are two main types of electronic computer, *analogue* and *digital* computers. The basic difference between them is the way in which they handle numbers for purposes of calculation. Analogue computers make use of variable quantities such as length, shaft rotation or different voltages. Digital computers, on the other hand, carry out their calculations solely with integers.

Analogue computers are mainly used in connection with the solution of problems concerning quantities which vary with time, and are often designed with specific problems in mind. Digital computers are general purpose machines which will do complex arithmetic at high speeds. Thus it can be seen that the two types are complementary. The analogue computer is used mainly for research and engineering design work. It is not, in comparison with the digital computer, often a high precision machine, since it is mainly concerned with varying quantities. The digital machine is used in research applications where great accuracy is required and in commercial applications where, for example, precision of one penny in several million pounds is needed.

Digital Computers

The digital computer does arithmetic at high speed under the control of orders which are given to it by its operator. The problem which the computer is to solve is first reduced to a series of orders which is known as the *programme*. The programme might be very simple, for example add *a* to *b*, divide the result by *c*, and add the result of this to *d*. Reducing problems to a series of orders such as this is called programming. Once the computer has been given the data—the numbers it is to add, subtract, multiply, etc.—and the programme, it computes the answer without any further action by the operator.

The essential parts of a digital computer are: (1) An input arrangement to enable the data which is to be calculated and the instructions for the control of the machine's operation—the programme—to be fed into the computer. (2) An output device to display the result of the calculation. (3) A storage unit, or *memory*, for holding the data being processed and the programme during the

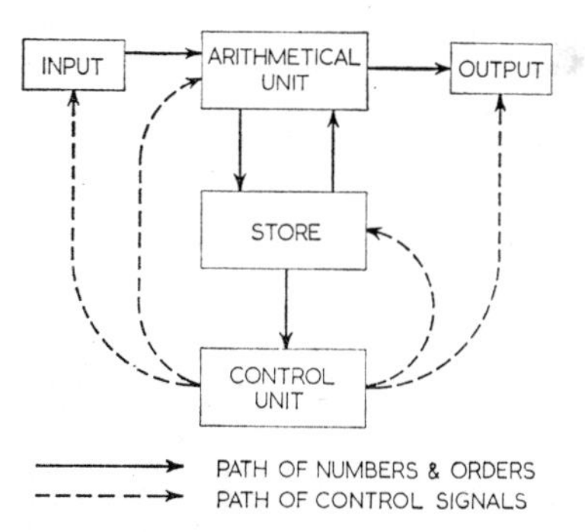

Fig. 13. Block diagram showing the basic sections of a digital computer and the flow of information in the computer between the various sections.

computation. (4) An arithmetic unit for actually carrying out the arithmetical operations. (5) A control unit which interprets the programme and controls the flow of data between the various parts of the computer. Fig. 13 shows in block diagram form the basic sections of a digital computer and the flow of data in it.

Binary notation

Most everyday calculations are carried out in decimal notation, using the figures 0 to 9. The components which perform the arithmetical operations in an electronic digital computer have, however, only two states (as opposed, for example, to the dekatron—see Chapter 3—which has ten), namely 'on' or 'off'. The diodes, valves, transistors or relays used in the arithmetic unit, that is, are either on, i.e. conducting, or off, non-conducting. It is therefore necessary in digital computers to use a system of notation based on two figures only. The binary system uses only the figures 0 and 1, which can be regarded by the computer's calculating components as either on or off.

Conversion from decimal notation to binary notation is done by repeatedly dividing the decimal number by 2 and indicating the remainder at each stage by 0 or 1 as the case may be. For example

Table 1. Decimal and Binary Numbers

Decimal number	*Conversion*		*Binary number*
0	0	0	0000
1	1	2^0	0001
2	2	2^1	0010
3	2 + 1	$2^1 + 2^0$	0011
4	4	2^2	0100
5	4 + 1	$2^2 + 2^0$	0101
6	4 + 2	$2^2 + 2^1$	0110
7	4 + 2 + 1	$2^2 + 2^1 + 2^2$	0111
8	8	2^3	1000
9	8 + 1	$2^3 + 2^0$	1001
10	8 + 2	$2^3 + 2^1$	1010
11	8 + 2 + 1	$2^3 + 2^1 + 2^\circ$	1011

the binary equivalent of the decimal number 75 is obtained as follows:

75	37	18	9	4	2	1	0
	1	1	0	1	0	0	1

The binary equivalent number, reading from right to left, is thus 1001011. Each digit indicates whether or not successive powers of 2 are present or absent in the number. Table 1 gives the binary equivalents of the decimal numbers 0 to 11.

Digital computer components

In the arithmetic unit the binary numbers are represented by circuits which are either in the 0 or 1 state (i.e. on or off). A series of these circuits will give any number and they will change very rapidly—in less than a millionth of a second—so that very fast calculating times are achieved.

In the memory unit it is usual to use large numbers of ferrite magnetic cores. These are also two state devices. If a current is passed through one it takes a magnetic state which can be made to represent say 0. When the current ceases it retains this state, thus acting as a memory. If a current is passed through it in the opposite direction, however, the magnetic state is reversed, and this reversed state can be made to represent the other figure, in this case 1.

Analogue computers

The analogue computer is used where a certain number of variable processes are to be simulated and the results of various interactions studied. Thus the analogue computer as a laboratory tool can simulate experiments and provide answers to problems which would take scientists many years to obtain by normal methods.

The accuracy of an analogue computer is mainly determined by the accuracy with which the physical factors that it is being used to study can be measured. The input to an analogue computer is often the output of transducers making measurements on a process under investigation, although it may also of course be a programme of some sort.

In the analogue computer, the input signals are applied to

amplifiers which, mainly by the application of suitable feedback networks (see Chapter 4), provide an output which consists of the solution to a particular algebraic equation, i.e. the relation between the input and output of the amplifier is the solution to a particular equation. Circuits can be devised which carry out such processes as summation, integration, differentiation etc.

Thus extremely complex problems, often varying with time, can be resolved by the use of an analogue computer.

II

THE FUTURE OF ELECTRONICS

We are today seeing the start of a new era in instrumentation and the end of one that began sixty years ago with the development of the thermionic valve. This new beginning started when semiconductor devices began to replace the thermionic valve.

There are three important tasks facing those who are leading electronics forward to make them play an even more significant role in human affairs. They are the reliability, physical size and performance of electronic devices.

Considerable advances have already been made in increasing the reliability of electronic equipment. In particular the use of semiconductor materials and devices has enabled equipment to be designed which can be operated without maintenance except of a routine nature for much longer periods than was previously possible.

Reduction in the size of components is an all-important aim in electronics. One of the reasons the computer cannot imitate the human brain is that enough components can not be assembled in a reasonable volume to allow sufficient inter-circuit links to be made.

The human brain has millions of circuits in the guise of neurons or brain cells. Every person from the age of thirty onwards loses brain cells at the rate of several thousands per day and yet their brains are still capable of intricate mental tasks. Imagine a computer in which several thousand transistors failed every day and yet it was able to continue with its allotted task! This may well be possible in the future: basic circuits such as amplifiers, pulse generators, switching circuits and so on have already been miniaturized, so that it will soon be possible to get complete circuits into a very small volume. It has been suggested that by using microminiaturization techniques it will be possible to get four million components into a cubic foot.

Higher performance usually means, to the electronics engineer, a higher signal-to-noise ratio. Electrical noise is always present everywhere and it is no use searching for and trying to rid oneself of it like one gets rid of flies or a wasp's nest. The most the engineer can do is to shield his equipment from as much noise as possible and design his equipment so that it generates as little internal noise as possible. Noise is not, however, a problem in all electronic undertakings. In the digital computer, for example, noise is relatively unimportant—unless you count the drift of transistors due to temperature variations which can, if large enough, produce an error signal and therefore can be classified as noise.

Noise

The term noise in electronics means any unwanted signals that interfere with or obscure the signal with which the equipment is primarily concerned. In Fig. 1 (a) a radar 'A' scan with a lot of

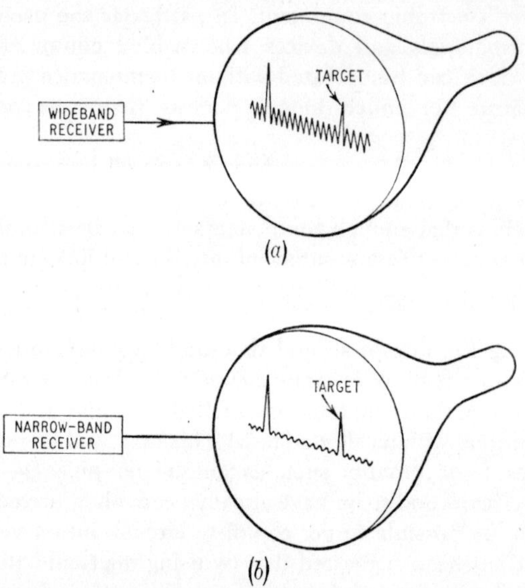

Fig. 1. Effect of noise on a radar display. The use of a narrow-band receiver reduces the extent to which the noise masks the signal obtained from the target.

noise is shown. A wideband receiver will allow a lot of background noise through because it is not very selective. In (b) the same signal is shown with the noise reduced by using a selective, narrow-band receiver. It is not always possible to get rid of noise as easily as this and in any case the receiving equipment can generate enough noise in its input stage to 'drown' a weak signal. Examples of very weak signals are the incoming signals from communication satellites and the tiny absorption signal from a minute piece of organic material containing free radicals in an electron spin resonator (see Chapter 7).

The noise generated in equipment is mainly caused by thermal agitation of electrons, especially in the resistors in the first stage of an amplifier or receiver. The electrons 'jiggling' about in a random fashion in a resistor behave as though they were a lot of generators all packed together in a small space. If these are connected to the grid circuit of a valve amplifier they will appear as an even larger noise voltage at the output due to the amplification of the valve. This noise will be quite big compared to the noise voltage generated by the resistors in the next stage of amplification, so that noise in later stages can be ignored to some extent. It is the first stage which is the real trouble so that the engineer tries to design the first stage of a receiver or amplifier so that it will contribute as little noise as possible.

The external noise that enters the receiver or amplifier along with the wanted signal is generated by electrons moving around in a random fashion in the atmosphere (e.g. ionospheric noise). The hotter the atmosphere the greater the electrical noise generated—this is why storms and the sun affect reception. If you hold your arms out as shown in Fig. 2 there will exist between your two hands a high resistance formed by the air between them. Since this is a high resistance a noise voltage is generated between these two points. If the hands are moved farther apart the noise voltage increases, and if a fire is lit underneath your hands the noise voltage would increase even more. Of course in normal life the presence of this noise voltage does not bother us, but, if your hands were the aerial of a sensitive receiver, this noise voltage would be picked up in addition to the wanted signal.

The signals from distant stars are very small and masked by noise voltages which make them very difficult to identify and pick up. Stations such as Jodrell Bank use huge aerial dishes to produce a highly concentrated polar response—only signals travelling in a very narrow beam are picked up. In this way noise signals outside the beam are excluded. A simple experiment will show you the

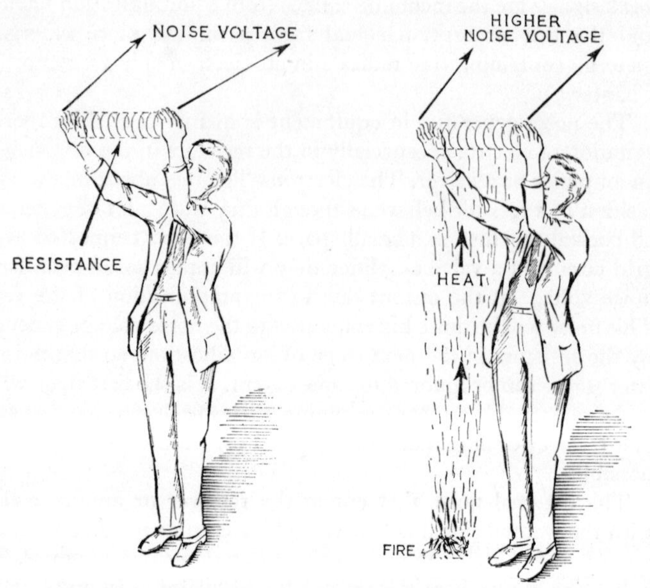

Fig. 2. Electrical noise in the atmosphere. Heat increases the noise present by agitating the electrons.

effectiveness of this method for yourself. Take a cardboard tube and study a distant object by looking at the object through the tube. Then look at the object in the normal way, dispensing with the aid of the tube. The object appears to be less clearly defined without the cardboard tube. This is because by using the tube a lot of unwanted light is excluded, vision being concentrated on the 'light signal'—the distant object. The same test can be made by listening with the aid of a long tube. If the tube is pointed at someone talking a long way off you will be surprised at the amount that you can hear. This method, using a microphone at one end of the tube, was in fact used in the U.S.A. to catch a spy who was talking to an associate at a distance of over a hundred yards from the microphone.

The battle against noise is one of the most important facing the future designer of electronic equipment.

Reducing noise

The only way that noise can be reduced in an electronic system is to use a sensitive, noise free amplifier as the first stage of the detecting apparatus. To achieve this aim a number of radically new amplifiers, which are very sensitive and can be used at very high frequencies, i.e. in the microwave region, are being developed. It is in the microwave region that a lot of current research work in electronics is being done. These amplifiers are mainly being developed for use in the communications field but are capable of being used in other fields.

The parametric amplifier

The performance of an amplifier circuit such as that shown in Fig. 3 is mainly limited by the characteristics of the valve and the 'goodness' (Q) of the anode circuit, which is tuned to the signal it is amplifying. The anode circuit consists of an inductance coil and capacitor in parallel: at the frequency to be amplified this combination behaves as though it is a large resistor (parallel resonance

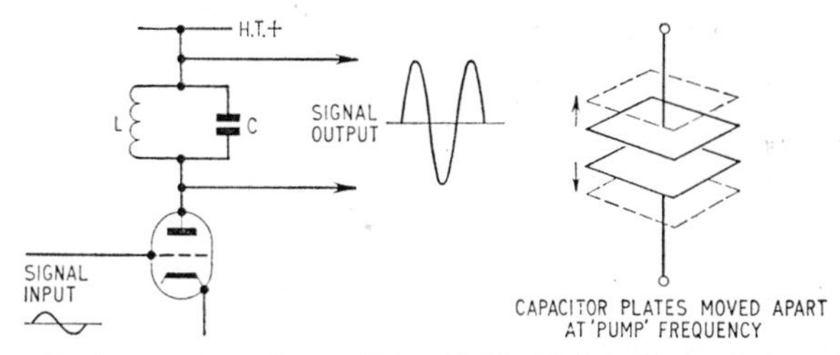

Fig. 3. Basic tuned anode amplifier (left). If the plates of capacitor C could be moved apart at a frequency equal to twice the frequency at which the amplifier is working the output would be increased. This is the principle of the parametric amplifier.

circuit) whilst at any other frequency it behaves as a very small resistor so that virtually no amplification occurs because no signal voltage is developed across it.

One way of achieving the best results from an amplifier of this type is to design it so that the energy loss of the inductor and capacitor is as small as possible. Nevertheless, however well this is done there is always some loss, which sets a limit to the performance of the circuit.

If, however, it was possible to take hold of the capacitor plates and pull them apart at a frequency equal to twice the frequency at which the amplifier is working then energy would be introduced into the circuit by the very act of pulling the plates apart. The energy expended in pulling the plates apart adds itself to the amplified signal in such a way as to increase the output.

This phenomenon can be observed with a child's swing. If the child is swinging back and forth with the child sitting perfectly still then the swing will gradually come to a halt due to the losses (friction) incurring at the point where the swing is fastened. If, however, the child moves up and down in the seat so as to effectively change the length of the swing rope at a frequency equal to twice the frequency he wishes to swing at, then the swing will move at this frequency and continue to build up in amplitude. Of course the child on the swing knows nothing of this: he simply acts intuitively.

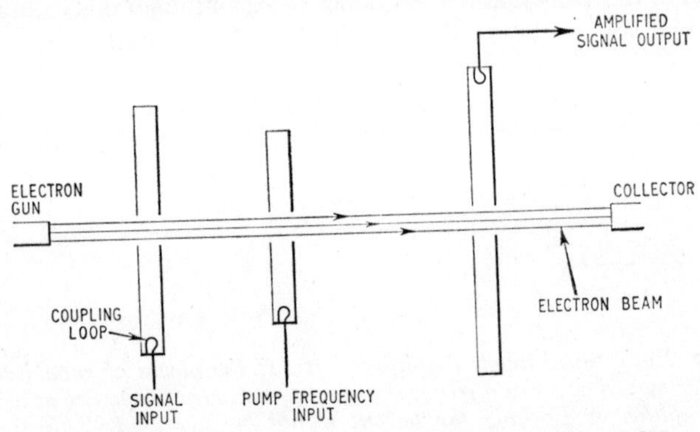

Fig. 4. The microwave electron beam parametric amplifier.

This principle has been used in an amplifier working in the microwave region called the *parametric amplifier*, the frequency at which the capacitor plates are moved being called the *pump frequency*. Fig. 4 shows an electron beam parametric amplifier. The electron beam couples the various 'electrodes', which are resonant cavities made of waveguides. The waveguide configuration also reduces the effects of the noise that occurs within the tube. Thus the parametric amplifier is a sensitive, low-noise device which offers a great deal of promise for the future. The signal is fed into the first cavity and coupled to the output by the electron beam, which is modulated by the signal. The pump frequency is introduced so that it does effectively the same job as in a conventional circuit, that is adds energy to the original signal.

The maser

The maser amplifier makes use of the energy stored in molecules called paramagnetic ions. This energy can be released when the correct r.f. and d.c. magnetic fields (supplied by superconductor magnets) are applied in the correct orientation to the material containing the paramagnetic ions.

The maser consists of a quantity of material such as ruby, a large magnet, some sections of waveguide and a means of cooling the ruby to a temperature several hundred degrees below zero.

These amplifiers are difficult to explain simply and so are best accepted for what they are: really they belong to the province of the physicist rather than that of the electronics engineer, who is only interested in their performance, which is very high indeed. Their use in receiving signals from satellites and in similar applications may be amongst some of the most important work in the future.

Microminiaturization

The trend for many years has been towards smaller components and the introduction of transistors to replace valves in many applications gave emphasis to the need for smaller components to use with them. Using printed circuit techniques and miniature and sub-miniature components it is possible to build equipment that will occupy only a fraction of the volume hitherto needed. For example, a digital computer which ten years ago would have occupied a small room can be be built into a box no bigger than an ordinary suitcase. Analogue computers, on the other hand, are more difficult to reduce in size. Any small drift in circuit values due to changes in ambient temperature will be regarded by an analogue computer as a new signal. In digital equipment drift has to be large before it affects the equipment. This difference between digital and analogue techniques can be illustrated by considering the speedometer and milometer of a car. The speedometer is an analogue device and the milometer a digital device. If any error occurs in a speedometer it will be immediately apparent on the dial. A false reading will be given to the driver and his estimate of his speed will be wrong. In the case of the milometer, however, small errors (as long as they are not big enough to equal a mile) will not even be indicated. At the end of the next mile, when the meter is about to register, the errors will have evened out and only a true mile will be registered.

Because transistors are temperature sensitive they were for a while not used in analogue computers, which could thus not be reduced much in size. However, to prevent transistors from drifting with temperature changes and giving false readings in an analogue equipment it is possible to house transistorized analogue equipment in a temperature-controlled cabinet. This technique is called *environmental stabilization* and is illustrated in Fig. 5. A transistorized analogue computer can now be built to fit a space the size of a coat locker instead of needing the space of a large room as was the case when valves were used.

Further size reduction in components and equipments can be achieved by a radical change in construction techniques. Three approaches to microminiaturization are possible: micromodules, microcircuits and solid state circuits.

The micromodule, developed mainly for missile applications, consists of a stack of ceramic wafers about a quarter of an inch square and 20 'thou' thick, held in a cage of riser-wires. Each wafer carries one or more components.

Microcircuits differ from micromodules in that the whole circuit is laid on one wafer. In this application the wafer may be up to one inch square.

The solid state circuit is a functional unit fashioned from a tiny block of semiconductor material. Using a silicon substrate any component except an inductor can be included in a single block of the material. The method of doing this is not too difficult and a

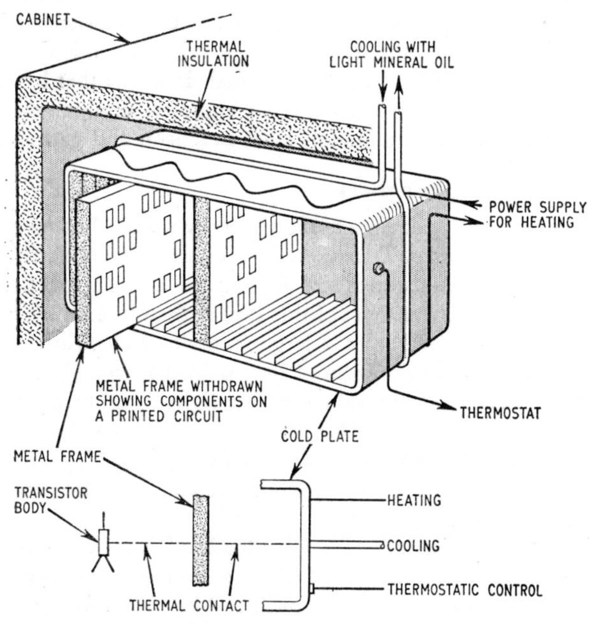

Fig. 5. Technique of environmental stabilization.

number of successful circuits have been produced. By using this technique several million components can be packed into a cubic foot.

The production of a solid state circuit is fascinating. If a semiconductor diode is reverse-biased—rather like the zener diode described in Chapter 3—a slightly different effect occurs due to a deliberate change introduced into the junction of the p and n type materials. Because of this the device acts as a capacitor. Thus it is possible to fashion both transistors and capacitors out of the same block of silicon. The resistor is simply a resistive path in the silicon substrate, which can be arranged to have any dimensions.

From this it becomes obvious that semiconducting materials may decide the future of much work in electronics. The different effects in semiconductor material—zener diode, capacitor, transistor, ordinary diode and so on—depend on the characteristics of the junction or junctions, the doping of the original semiconducting materials and the external voltages applied. It seems that by using a variety of manufacturing techniques complete equipments may be made up using only materials such as silicon and germanium.

Super conducting magnets

Conventional electromagnets are generally made of copper coils wound on an iron yoke. The production of large magnetic fields in such an arrangement requires a large yoke and high current density in the coils. An electromagnet of this type is therefore bulky, heavy and costly. It also consumes a large amount of electrical energy which is dissipated in the form of heat so that forced air or water cooling is necessary. A typical example weighs roughly 3 tons and is powered by a 20-kW generator to provide magnetic field strengths up to 20 kilogauss in a 2 in. gap between the poles. Magnetic field strengths of this order are becoming more and more important in electronic research.

If the coils are wound with wire made of a superconducting material such as niobium instead of the conventional copper much smaller magnets of the same strength can be made. When in the superconducting state, a material such as niobium loses all traces of electrical resistance and can thus carry large electric currents without producing heat. This makes possible the design of a superconducting magnet which will carry larger current densities than could be employed with copper coils, with consequent reduction in size and weight. A coil wound with niobium-zirconium alloy wire capable of producing fields greater than 50 kilogauss would weigh only a few pounds. The property of superconductivity is dependent on temperature. It usually occurs at about 7° K (−266°C).

High-speed camera

Another interesting development is the high-speed microsecond camera. A mirror rotating at 6,000 revolutions per second sweeps past a series of lenses held in an arc of about the same shape and size as a motor car brake lining. Close to these is another arc of photosensitive paper or film. The beam from the mirror traverses the paper, as shown in Fig. 6, producing in it a series of pictures

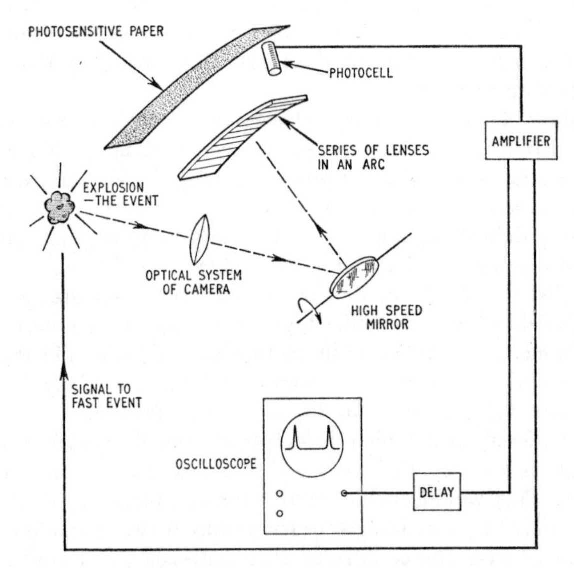

Fig. 6. Electronically controlled high-speed camera.

taken at high speed. This technique is useful in photographing some event such as a nuclear explosion or some experimental work, for example breaking down an insulator with a high current.

The camera and the subject being photographed are synchronized by means of an electronic system comprising amplifiers and delays initiated by a photocell. When the light beam strikes the photocell it produces a signal which starts the event and also starts a cathode-ray tube trace. The camera is able to photograph events which hitherto had defied analysis.

Other developments

In addition to the developments already mentioned there is a great deal of 'bread and butter' work going on to improve the results obtained from existing equipments and systems.

The increase from 405 lines to 625 lines for the television line standard means improved definition pictures, and the introduction of colour television may produce a startling change in the viewing habits of most people.

Colour television, using closed-circuit techniques (i.e. the information is sent from camera to receiver along a cable) is an aid to education, especially in medical teaching. In the U.S.A. a number of these systems are in use in medical schools to enable students to study delicate operations. At the same time they are given an expert commentary by a surgeon-teacher.

Whilst on the subject of medicine, many new ideas are being tried in this field which may, indeed, prove to be one of the most fruitful for the electronics engineer in the future. The radio pill has already been mentioned in an earlier chapter and it is true to say that it has still to be fully exploited by the medical profession.

A national grid which links people together with television circuits is another idea envisaged at a recent conference on television. This would enable many large companies to decentralize their offices by providing television links between satellite offices spread all over the country so that materials and layout may be disposed in a more economical manner.

The reduction in the size of computers and their subsequent lower cost will mean a great increase in their use over the next decade and many of the more tedious jobs in industry will be done by equipment under the control of electronic 'brains'.

12

TRAINING TO BE AN ELECTRONICS ENGINEER

n the early days of wireless it was possible for anyone who wished o do so to become a wireless (electronics) engineer. Today, how- ver, there is less chance of reaching the top without some formal raining.

In recent years the members of this profession have come to be livided into four groups: engineer, technician, craftsman and perative. The distinctions between these groups seem clear enough n paper but are difficult to define in practice: there are cases where man who is no more qualified on paper than the craftsman is the eader of a group doing advanced work in electronics.

For the sake of clarity the following definitions of the various rades are given.

The professional engineer is one who is competent by his train- ng to apply scientific method to the analysis and solution of echnological problems in his own field. This means that faced with ny problem of a technological nature he can plan, organize and lecide how the problem should be tackled. He is aware of the ossibilities of the materials and instruments and techniques at his ommand and can marshal them to solve his problem.

The technician, on the other hand, is one who is qualified by irtue of his education to apply in a responsible manner those echniques prescribed by the professional engineer. The technician s the sort of person who when faced with a difficulty during the onstruction of a new system devised by the engineer is capable of electing the correct components, supervising the construction, esting and calibrating of the equipment.

The craftsman must be capable of wiring circuits neatly and in ccordance with the latest method, and must be able to read ircuit diagrams.

The operative is someone who is capable of carrying out simple operations under appropriate supervision—such things as simple measurements or fitting components to an equipment.

Qualifications needed

An operative is more than likely today to be a woman or girl who could just as easily be working at some other factory job in the area where she lives, although a certain manual dexterity is needed to work on electronic equipment.

To be a craftsman requires experience and an interest in electronics. Beyond intelligence there is little need for scholastic ability although simple calculations are occasionally required of the craftsman. The craftsman should be good with his hands: he is a type of person who cannot wait to acquire a lot of seemingly useless knowledge but must be getting on with things—he likes the feel of materials and tools. This attitude coupled with an interest in electronics may be a useful beginning to a long and happy career.

The technician often has to do a job which is at least as difficult as the engineer's and his educational training may not fall far short of that of an engineer. Like the craftsman he is impatient to begin and so usually starts work before he is fully trained, although deficiencies in basic knowledge can usually be made up for by attendance at evening classes. The technician usually qualifies by taking all or some of either the City and Guilds of London Institute examinations in the appropriate subjects or the National Certificate examinations. In practice there are many technicians who have not taken any of these qualifications at all and are yet very efficient at their jobs. Many who have taken all their examinations, for instance a C.G.L.I. full technological certificate or a Higher National Certificate, may find themselves treated as professional engineers although normally they are required to take some extra examinations to qualify for Associate Membership of the appropriate Professional Institute.

The best advice that can be given to a young person entering the profession is to do as well as he can with the facilities available to him: read books and ask questions. One thing is certain in this profession—the ability to do the job is more important than paper qualifications.

How to qualify

The best qualified engineer is perhaps the university graduate. Leaving a grammar or public school with about seven 'O' level passes and three suitable 'A' level passes in the G.C.E. examination he goes on to university and takes a degree in electrical engineering —specializing in 'light current' techniques. After leaving university he might join one of the large industrial firms in the field or go into the Government Scientific Service, doing a two-year course as a graduate apprentice. During this time he will move from job to job within the organization, learning all the time. He may take some evening post-graduate course at a local technical college in the subject nearest to his job—for example he may study computers, or aerial design, guidance systems or something along these lines.

After two years he should find himself promoted to full engineer status and since he will now be about 23 years of age he may expect to receive a salary in the region of £1,000 per annum.

After several years in the post of his choice, he will probably have assumed some responsibility—perhaps as a development engineer, in charge of the installation of equipment or on the sales or management side or in research. His salary at the age of thirty might be about £1,500 rising to £2,000 per annum or higher. He could then apply to his Professional Institute, for example the British Institution of Radio Engineers, for corporate membership and become, say, John Smith, Esq., B.Sc., A.M.Brit.I.R.E., Chartered Engineer.

This is the theoretical case but it should be remembered that there are many men in industry earning high salaries and proving themselves to be first class engineers who rose from the workbench and have no formal qualifications at all.

Another route to professional qualification is the student-apprenticeship. A boy (or girl) may leave school with say two 'A' levels and the usual number of 'O' levels in the G.C.E. examinations. This is not sufficient to get him a place at most universities and so application could be made for a student apprenticeship with a large electronics firm. At the age of seventeen or eighteen years the student apprentice begins a sandwich course. This means that for several years he spends alternatively six months in industry and six months at a Technical College or similar establishment. All this time the student receives a salary from his firm. At the end of four or five years the student apprentice sits the examination for the Diploma in Technology, the Dip.Tech. This is a qualification which will lead to professional status and Associate Membership of the appropriate Professional Institution.

Perhaps the most arduous way to qualify is to struggle up the ladder unaided. Such a person may have started off on the wrong foot; perhaps he failed the eleven plus. At the age of fifteen he leaves school without any definite ideas as to his future but with a vague interest in electronics. He may then get a job as an operative or craft apprentice in a radio factory, or he may join the forces to learn the trade of radio or radar mechanic. If he is keen to learn, reads as many books as he can and shows an interest in the subject he will seldom fail to find encouragement offered to him. He can enrol at a local Technical College and study either for his C.G.L.I. Intermediate Certificate or else his Ordinary National Certificate—it is advisable to take whichever qualification the college appears to concentrate most upon.

After two or three years at evening classes he can sit the appropriate examination and if he passes he may find the world a different place. He might find himself promoted to a student apprenticeship by his firm; even if this does not happen he will very likely be given greater opportunities and a chance to prove his worth.

From this point on his chances are good. He may choose to continue studying and sit for the Higher National or C.G.L.I. Full Technological Certificate. After obtaining one of these higher qualifications he may decide to stop and become a technician. He may expect to receive a salary of about £1,400 at the age of 35 years and if he shows enough ability he could well become Chief Engineer of his Company.

INDEX